Advances in Physics of Rydberg Atoms and Molecules

Advances in Physics of Rydberg Atoms and Molecules

Eugene Oks
Department of Physics, Auburn University, Auburn, USA

IOP Publishing, Bristol, UK

ISBN 978-0-7503-3939-1 (ebook)
ISBN 978-0-7503-3937-7 (print)
ISBN 978-0-7503-3940-7 (myPrint)
ISBN 978-0-7503-3938-4 (mobi)

DOI 10.1088/978-0-7503-3939-1

Version: 20211101

IOP ebooks

British Library Cataloguing-in-Publication Data: A catalogue record for this book is available from the British Library.

Published by IOP Publishing, wholly owned by The Institute of Physics, London

IOP Publishing, Temple Circus, Temple Way, Bristol, BS1 6HG, UK

US Office: IOP Publishing, Inc., 190 North Independence Mall West, Suite 601, Philadelphia, PA 19106, USA

To Rjonis

Contents

Author biography

Eugene Oks

Eugene Oks received his PhD degree from the Moscow Institute of Physics and Technology, and later the highest degree of Doctor of Sciences from the Institute of General Physics of the Academy of Sciences of the USSR by the decision of the Scientific Council led by the Nobel Prize winner, academician A M Prokhorov. According to the Statute of the Doctor of Sciences degree, this highest degree is awarded only to the most outstanding PhD scientists who founded a new research field of great interest. Oks worked in Moscow (USSR) as the head of a research unit at the Center for Studying Surfaces and Vacuum, then at the Ruhr University in Bochum (Germany) as an invited professor, and for the last 30 years at the Physics Department of the Auburn University (USA) in the position of Professor. He conducted research in five areas: atomic and molecular physics, astrophysics, plasma physics, laser physics, and nonlinear dynamics. He founded/co-founded and developed new research fields, such as intra-Stark spectroscopy (new class of nonlinear optical phenomena in plasmas), masing without inversion (advanced schemes for generating/ amplifying coherent microwave radiation), and quantum chaos (nonlinear dynamics in the microscopic world). He also developed a large number of advanced spectroscopic methods for diagnosing various laboratory and astrophysical plasmas—the methods that were then used and are used by many experimental groups around the world. He recently revealed that there are two flavors of hydrogen atoms, as proven by the analysis of atomic experiments; there is also a possible astrophysical proof—from observations of the 21 cm radio line from the early Universe and from the observed distribution of dark matter in the Universe. He showed that dark matter or at least a part of it can be represented by the second flavor of hydrogen atoms. He has published about 500 papers and 9 books, including the books 'Plasma Spectroscopy: The Influence of Microwave and Laser Fields', 'Stark Broadening of Hydrogen and Hydrogenlike Spectral Lines in Plasmas: The Physical Insight', 'Breaking Paradigms in Atomic and Molecular Physics', 'Diagnostics of Laboratory and Astrophysical Plasmas Using Spectral Lineshapes of One-, Two, and Three-Electron Systems', 'Unexpected Similarities of the Universe with Atomic and Molecular Systems: What a Beautiful World', 'Analytical Advances in Quantum and Celestial Mechanics: Separating Rapid and Slow Subsystems', 'Advances in X-Ray Spectroscopy of Laser Plasmas', and 'Simple Atomic and Molecular Systems: New Results and Applications'. He is the Chief Editor of the journal *International Review of Atomic and Molecular Physics* and the Editor-in-Chief of the Physical Sciences section of the journal *Foundations*. He is a member of the Editorial Boards of six other journals: *Symmetry*, *American Journal of Astronomy and Astrophysics*, *Dynamics*, *Open Journal of Microphysics*, *Physics International* and *Open Physics*. He is a member of the Reviewers Board of the journal *Atoms*. He is also a member of the International Program Committees of the two series of conferences: *Spectral Line Shapes*, as well as *Zvenigorod Conference on Plasma Physics and Controlled Fusion*.

Chapter 1

Introduction

Rydberg (i.e., highly-excited) atoms and molecules have been studied for almost a century (we direct readers to relatively recent books, reviews, and papers [1–9] and references therein). This is because of their striking distinctions from atoms and molecules in the ground or a low-excited states. For Rydberg atoms, with the increase of the principal quantum number n, their characteristic size and the electric dipole moment scales as n^2 (which is why they are extremely sensitive to even weak electric fields and electromagnetic radiation) and their radiative lifetime scales as n^3. They strongly interact not only with an electric field, but also with each other: their dipole–dipole interaction scales as n^4/R^3 and their van der Waals interaction scales as n^{11}/R^6, where R is their separation from each other.

In laboratory conditions experimentalists deal with Rydberg atoms having the quantum number of several tens. In the Universe, in H II regions, the principal quantum number of Rydberg atoms can reach $n \sim 1000$. In the latter case, the Rydberg atoms cannot be considered anymore as microscopic objects: the diameter of their electron orbit is ~0.1 mm, i.e., twice the diameter of human hair.

Studies of Rydberg atoms and molecules have numerous applications. Examples are highly-responsive sensing, quantum optics (where the light is manipulated at the level of a single photon), quantum computing, quantum chaos, ultracold Rydberg gases and plasmas, anti-hydrogen formation, and plasma diagnostics.

The most recent book on Rydberg systems—the good, but relatively small electronic book of only 27 pages by Šibalić and Adams published in 2018 [1]—focused mostly at the technological applications and also was limited to Rydberg atoms, i.e., did not discuss Rydberg molecules. In distinction, in the present book we focus on the theoretical advances in physics of Rydberg systems and include in the consideration both Rydberg atoms and Rydberg molecules. Besides, the present book is almost five times larger in size than their book [1].

The book is structured as follows. In chapter 2 we follow paper [10] to present the study of classical circular Rydberg states of a hydrogenic atom/ion under collinear electric and magnetic fields of arbitrary strengths, the entire system being immersed into a plasma with the Debye screening. We show in detail how the screening decreases the value of the critical electric field required for the ionization at different values of the magnetic field. The results should have fundamental importance because hydrogenic atoms/ions under external fields remain a test-bench for atomic physics. Also the results could motivate experiments on the magnetic control of the 'continuum lowering' in cold Rydberg plasmas, this being of practical importance.

It should be emphasized that there have been lots of theoretical and experimental studies of circular Rydberg states (see, e.g., papers [11–14] and references therein). The reasons are as follows. First, their properties facilitated works on inhibited spontaneous emission and cold Rydberg gases (see, e.g., papers [15–17]). Second, classical circular Rydberg states are counterparts of quantal coherent states. Third, in the quantal method using the expansion in terms of the inverse value of the principal quantum number, classical circular Rydberg states are the primary term (see, e.g. paper [18] and references therein).

In chapter 3 we follow paper [19] to present a study of an isolated two-electron Rydberg atom or ion of the nuclear charge Z, where the average distance of the inner electron from the nucleus is much smaller than the average distance of the outer, Rydberg electron from the nucleus. In this situation, the orbit of the outer electron is very close to the Kepler ellipse of an electron in a hydrogenic atom or ion of the effective nuclear charge (Z−1). Therefore, the hydrogenic subsystem Z+e, consisting of the nucleus Z and the inner electron, is under the rotating electric field whose magnitude and the angular velocity vary in time. In paper [19] the author obtained analytically the shape of any spectral line emitted by the subsystem Z+e in this situation. As a particular example, he studied in detail the shape of the Ly-alpha line. The obtained analytical expressions yielded, in particular, a *peculiar result*. Namely, as the spectral line splits into several components, the most intense components exhibit a *quadratic* Stark effect with respect to the electric field of the outer electron (though it is linear with respect to the dipole moment of the inner electron), while one would intuitively expect a linear dependence on the electric field of the outer electron since the subsystem Z+e is hydrogenic. Physically this counterintuitive result is due to the rotating nature of the electric field of the outer electron, as explained in detail in chapter 3.

Chapter 4 presents the study of a helium atom or a helium-like ion with one of the two electrons in a Rydberg state, the system being under a high-frequency laser field. For obtaining analytical results, we use the generalized method of the effective potentials. We find two primary effects of the high-frequency laser field on circular Rydberg states. The first effect is the precession of the orbital plane of the Rydberg electron. We calculate analytically the precession frequency and show that it differs from the case of a hydrogenic atom/ion. In the radiation spectrum this precession would manifest as satellites separated from the spectral line at the Kepler frequency by multiples of the precession frequency. The second effect is a shift of the energy of the Rydberg electron, also calculated analytically. We find that the absolute value of the shift increases monotonically as the unperturbed binding energy of the Rydberg electron increases. We also find that the shift has a non-monotonic dependence on the nuclear charge Z: as Z increases, the absolute value of the shift first increases, then reaches a maximum, and then decreases. The non-monotonic dependence of the laser-field-caused energy shift on the nuclear charge is a *counterintuitive* result.

Chapter 5 is devoted to negative hydrogen ions, such as, H^- and D^-. They are important from both practical and fundamental points of view. One of the reasons for their practical importance has to do with heating tokamak plasmas by external beams of negative hydrogen ions. Another practical use is the injection of negative hydrogen ions in high power proton accelerator facilities—for enhancing their performance. In addition, negative hydrogen ions are also important for studies of opacities of the atmospheres of the Sun and of the A-type stars. In this chapter we present a classical analytical description of negative hydrogen ions in the situation where the underlying hydrogen (or deuterium) atom constitutes a rapid subsystem while the outer electron represents a slow subsystem. We focus on the case where the inner electron is in a circular state, so that the subsystem 'nucleus plus inner electron' does not have the average electric dipole moment—in distinction to previous studies where the presence of the average dipole moment was the crucial requirement. By using the separation of rapid and slow subsystems, we show analytically that there is a classical bound state in such system and study its parameters. In particular, we calculate analytically the primary frequency of the radiation of such a system. This could be used for its experimental detection. The states that we found for the above systems could be considered as classical counterparts of the double-excited states of negative hydrogen ions previously studied in the literature in frames of quantum mechanics. The existence of classical counterparts of the double-excited states of negative hydrogen ions is a *counterintuitive* result.

Chapter 6 presents results of the study [20] dealing with the Stark broadening of hydrogen radio recombination lines, observed from galactic H II regions, the broadening being controlled by inelastic collisions with charged particles. There was a dramatic discrepancy—up to several hundred percent—between cross-sections of inelastic electronic collisions (causing the radiative transitions between the Rydberg levels) calculated by various authors. This dramatic discrepancy in cross-sections leads to a significant discrepancy in spectral line widths of radio recombination lines. In the study [20] the authors addressed two problems: (1) whose results for the cross-sections of the radiative transitions between the Rydberg levels are the most reliable and should be recommended to astrophysicists/observers for their analysis of the measured profiles of radio recombination lines? (2) How can one allow for all major processes relevant to the Stark broadening of radio recombination lines (transitions between discrete levels, ionization, charge exchange) within a single/unified theoretical approach? For this purpose the authors of paper [20] employed calculations in frames of the classical trajectory Monte Carlo approach. They obtained the results for a broad range of velocities of the charged particles (both electrons and protons), including the region most difficult for the analytical description, where the velocities of both the incident particle and the atomic electron have the same order of magnitude. By benchmarking analytical and semi-empirical results of other authors to the calculations from [20] (which are non-perturbative and do not use the dipole approximation), the authors found which one is the most accurate and should be recommended to astrophysicists/observers for their analysis of the measured profiles of radio recombination lines.

It should be emphasized that radio recombination lines of atomic hydrogen emitted at transitions between neighboring Rydberg energy levels were detected for the first time in the spectra of emission nebulae in 1964 [21, 22]. Since then, radio recombination lines have become a powerful tool for studying various processes in the nebulae, also known as H II regions (see, e.g., [23, 24]). An important role in developing the theory of the Stark broadening of radio recombination lines of atomic hydrogen had been played by works [25, 26]. The authors of these works showed that the contribution of elastic collisions with charged particles is considerably reduced by a near-compensation of perturbations of upper and lower levels. As a consequence, the main contribution to the Stark broadening of radio recombination lines comes from inelastic collisions with charged particles.

In chapter 7 we follow paper [27] to present a detailed classical description of the oscillatory-precessional motion of an electron in the field of an electric dipole. Specifically, we demonstrate that in the general case of the oscillatory-precessional motion of the electron, the oscillations being in the meridional direction (θ-direction) and the precession being along parallels of latitude (φ-direction), both the θ-oscillations and the φ-precessions can actually occur on the same time scale—contrary to the statement from the work by another author. We obtain the dependence of φ on θ the time evolution of the dynamical variable θ, the period T_θ of the θ-oscillations, and the change of the angular variable φ during one half-period of the θ-motion—all in the forms of one-fold integrals in the general case and illustrate it pictorially. We also produce the corresponding explicit analytical expressions for relatively small values of the projection p_φ of the angular momentum on the axis of the electric dipole. We also derive a general condition for this conditionally-periodic motion to become periodic (the trajectory of the electron would become a closed curve) and then provide examples of the values of p_φ for this to happen. Besides, for the particular case of $p_\varphi = 0$ we produce an explicit analytical result for the dependence of the time t on θ. For the opposite particular case, where p_φ is equal to its maximum possible value (consistent with the bound motion), we derive an explicit analytical result for the period of the revolution of the electron along the parallel of latitude.

Chapter 8 presents new possibilities for atomic physics experiments to reveal more detailed properties of nuclei. It is well-known that atomic physics experiments can reveal some properties of nuclei—see, e.g., review [28] and references therein. For example, observations of isotope shifts were used for determining nuclear sizes: the corresponding results obtained under the assumption of the spherical nuclear shapes for the uniform charge distribution can be found even in the textbooks on quantum mechanics—see, e.g., Flügge's book [29]. Further results, obtained under the assumption of the spherical nuclear shapes, but for Gaussian and the so-called exponential charge distributions, can be found in paper [30]. We note that the Lamb shift calculations are also usually performed under the assumption of the spherical nuclear shapes—see, e.g., paper [31] where such calculations were done for the Gaussian and Lorentzian (called 'dipole' in [31]) charge distributions.

However, many different atomic nuclei have non-spherical shapes, such as, e.g., (though not limited to) oblate or prolate ellipsoid of revolution (spheroid)—see, e.g., papers [32–34] and references therein. (For brevity we will use the term 'ellipsoid' for the ellipsoid of revolution.)

In chapter 8 we follow new results on this subject from paper [35]. We present analytical calculation of the shift of spectral lines of hydrogenlike ions for ellipsoid-shaped nuclei for the uniform charge distribution. In this situation the contribution to the shift originates not only from the part of the ionic ('atomic') wave function inside the nucleus, as was the case for spherical nuclei, but also from the modification of the ionic wave function outside the nucleus—in distinction to the case of spherical nuclei. (By ionic wave function we mean the wave function of the bound charged lepton, e.g., electron or muon.)

We show that the allowance for the ellipsoidal nuclear shape can change the shift of spectral lines of muonic ions by several times, compared to the corresponding shift for spherical nuclei. This can serve as an additional method for the experimental determination of the quadrupole moment of nuclei and of the standard beta-parameter related to the quadrupole moment.

In chapter 9 we present the study of the muonic-electronic negative hydrogen ion from our paper [36]. Investigations of muonic atoms and molecules are important for several reasons, such as the muon-catalyzed fusion (see, e.g., [37–39] and references therein), the laser-control of nuclear processes [40], the hunt for strongly interacting massive particles, once hypothesized as a possible dark matter candidate (see, e.g., [41] and references therein). Some other muonic-electronic systems were analyzed in more recent papers [42, 43].

As for investigations of the negative ion of hydrogen H^- (i.e., the *epe*-system), they are important because it is one of the simplest atomic systems exhibiting strong correlations, and because it plays a significant role in the D-layer of the Earth's atmosphere, in the atmosphere of stars (including the Sun), and in the design of particle accelerators [44].

In paper [36], which we overview here, the authors conjoined the above two research areas, by considering the μpe-system: the muonic-electronic negative hydrogen ion. They focused on the circular states. They demonstrated that the electronic motion can be considered as the slow subsystem, and the muonic motion as the rapid subsystem.

The authors of paper [36] showed that this physical system can behave as a quasimolecule where the muon rapidly rotates about the axis connecting the electron and the proton, while this axis slowly rotates. They also demonstrated that the muonic-emitted spectral lines undergo a red shift, which can be used for detecting such a system.

In chapter 10 we overview the results from paper [45] where the authors generalized the investigation from paper [36]. We analyze the quasimolecule μZe, where compared to the study from paper [36] the proton was substituted by a nucleus of charge Z.

In paper [45] it was demonstrated that for the validity of separating rapid and slow subsystems, the ratio of the angular momenta of the muon and the electron should slightly exceed the corresponding ratio from the case of $Z = 1$. It was also shown that for $Z > 1$, the muon binding energy significantly exceeds the corresponding binding energy from the case of $Z = 1$.

Finally, in paper [45] it was demonstrated that as the nuclear charge Z grows, the red shift of the muon-emitted spectral lines diminishes. However, it continues to be within the spectral resolution of typical spectrometers as a minimum till $Z = 5$, so that this red shift is still possible to observe.

In chapter 11 we overview the analytical results on one-electron Rydberg quasimolecules in a *circularly-polarized* high-frequency laser field from our paper [46]. This study complemented our previous investigation of one-electron Rydberg quasimolecules in a *linearly-polarized* high-frequency laser field presented in paper [47]. In paper [46], the following two major distinctions from the case of the linear polarization have been found

First, in the case of the linear polarization, as the laser amplitude grows, the number of the energy terms increases. However, in the case of the circular polarization no additional energy terms appear.

Second, in the case of the linear polarization, the electron radiation frequency shifted to the blue. However, in the case of the circular polarization, the electron radiation frequency shifted to the red. When the laser amplitude is known, one can perform the experimental determination of the separation between the nucleus and the electron orbital plane by measuring the frequency shift of the radiation.

In chapter 12 we present results obtained while we consider a muonic-electronic heliumic system in Rydberg states—starting from the concept from our paper [48]. There is a *counterintuitive* result, as follows.

While hydrogenic atoms/ion exhibit the linear Stark effect, helium atoms and heliumlike ions (hereafter, heliumic systems) do not demonstrate the linear Stark effect in a relatively weak external electric field—according to the existing paradigm (see, e.g., the textbooks [49, 50]). The overwhelming majority of states of hydrogenic systems have a non-zero value of the averaged electric dipole moment, which is the reason for the linear Stark effect in these systems, as is well-known. We recall that the averaging is performed in the quantum formalism over unperturbed wave functions or in the classical formalism (relevant to highly-excited states) over the unperturbed electronic orbit.

In distinction, according to the existing paradigm, for heliumic systems the averaged electric dipole moment is zero. Therefore, no linear Stark effect in a weak field should be exhibited by heliumic systems—even if one of the electrons is highly-excited. In the classical description, the unperturbed elliptical orbit of the highly-excited electron precesses. The precession is caused by the fact that at relatively small distances from the nucleus of charge Z, the effective potential deviates from the Coulomb potential of the effective nuclear charge $(Z-1)$. Of course, the unperturbed elliptical orbit of the electron would have a non-zero electric dipole moment; however, after averaging over the precession it becomes zero.

In paper [51], which we overview in this chapter, the author reconsidered the same physical system as in paper [48], but from a different point of view. Namely, he showed that the averaged electric dipole moment differs from zero in two states of the system. Therefore, the system in these states would respond even to a weak electric field by the linear Stark effect.

Besides, the author of paper [51] showed that by using the external electric field one can control the electronic orbit precession. He pointed out in particular that the precession can be stopped by applying a certain value of the external electric field and calculated this critical value.

The last section of chapter 12 is devoted to the following. As only circular orbits of the muon and the nucleus were assumed in paper [48], the open question was whether the results would significantly change by taking into account the ellipticity of their orbits. This question was addressed in paper [52], which we overview in this section.

For the motion of the electron, the authors of paper [52] obtained an extra term in the effective potential—the term depending on the eccentricity ε of the muon and nucleus orbits. For the case of the planar geometry, they demonstrated that, as the outcome, the frequency of the precession of the orbit of the electron acquires an extra contribution that depends on ε. They showed that the latter contribution, by the order of magnitude, can become comparable with the precession frequency for the case of the circular orbits of the muon and nucleus.

In appendices A and B we present details of calculations referred to from chapter 10 and from section 12.2, respectively.

References

[1] Šibalić N, Adams C S and Charles 2018 *Rydberg Physics Physics World Discovery* (Bristol: IOP Publishing)
[2] Browaeys A and Lahave T 2020 *Nat. Phys.* **16** 132
[3] Adams C S, Pritchard J D and Shaffer J P 2019 *J. Phys. B: At. Mol. Opt. Phys.* **53** 012002
[4] Cortiñas R G, Favier M, Ravon B, Méhaignerie P, Machu Y, Raimond J M, Sayrin C and Brune M 2020 *Phys. Rev. Lett.* **124** 123201
[5] Anderson D A, Sapiro R E and Raithel G 2020 *IEEE Aerosp. Electron. Syst. Mag.* **35** 48
[6] Letunov A and Lisitsa V 2020 *Universe* **6** 157
[7] Gnedin Y N, Mihajlov A A, Ignjatović L M, Sakan N M, Srećković V A, Zakharov M Y, Bezuglov N N and Klycharev A N 2009 *New Astron. Rev.* **53** 259
[8] Murray-Krezan J 2008 *Am. J. Phys.* **76** 1007
[9] Gallagher T 2006 *Am. J. Phys.* **63** 235
[10] Kryukov N and Oks E 2020 *Eur. Phys. J.* D **74** 125
[11] Lee E, Farrelly D and Uzer T 1997 *Opt. Express* **1** 221
[12] Germann T C, Herschbach D R, Dunn M and Watson D K 1995 *Phys. Rev. Lett.* **74** 658
[13] Cheng C H, Lee C Y and Gallagher T F 1994 *Phys. Rev. Lett.* **73** 3078
[14] Chen L, Cheret M, Roussel F and Spiess G 1993 *J. Phys.* B **26** L437
[15] Dutta S K, Feldbaum D, Walz-Flannigan A, Guest J R and Raithel G 2001 *Phys. Rev. Lett.* **86** 3993
[16] Hulet R G, Hilfer E S and Kleppner D 1985 *Phys. Rev. Lett.* **55** 2137
[17] MacAdam K B and Horsdal-Petersen E 2003 *J. Phys.* B **36** R167
[18] Vainberg V M, Popov V S and Sergeev A V 1990 *Sov. Phys. JETP* **71** 470
[19] Oks E 2017 *J. Phys. B: At. Mol. Opt. Phys.* **50** 115001

[20] Gavrilenko V P and Oks E 2007 *Phys. Scripta* **76** 43

[21] Dravskikh A F, Dravskikh Z V, Kolbasov V A, Misezhnikov G S, Nikulin D E and Shteinsheiger V B 1965 *Dokl. Akad. Nauk SSSR* **163** 332
Dravskikh A F, Dravskikh Z V, Kolbasov V A, Misezhnikov G S, Nikulin D E and Shteinsheiger V B 1966 *Sov. Phys. – Dokl.* **10** 627

[22] Sorochenko R L and Borodzich E V 1965 *Dokl. Akad. Nauk SSSR* **163** 603
Sorochenko R L and Borodzich E V 1966 *Sov. Phys. – Dokl.* **10** 588

[23] Gordon M A and Sorochenko R L 2002 *Radio Recombination Lines: Their Physics and Astronomical Applications* (Dordrecht: Kluwer)

[24] Mosse C, Calisti A, Stamm R, Talin B, Bureyeva L A and Lisitsa V S 2004 *J. Phys. B: At. Mol. Opt. Phys.* **37** 1343

[25] Griem H R 1967 *Astrophys. J.* **148** 547

[26] Minaeva L A and Sobelman I I 1968 *J. Quant. Spectr. Rad. Transfer* **8** 783

[27] Oks E 2020 *Symmetry* **12** 1275

[28] Neugart R 1992 Progress in atomic physics experiments on nuclear properties *Nuclear Shapes and Nuclear Structures at Low Excitation Energies* ed M Vergnes, J Sauvage, P-H Heenen and H T Duong (New York: Plenum) p62

[29] Flügge S 1974 *Practical Quantum Mechanics Part II* (Berlin: Springer) p 191

[30] Friar J L 1979 *Ann. Phys.* **122** 151

[31] Krutov A A, Martynenko A P, Martynenko F A and Sukhorukova O S 2016 *Phys. Rev.* A **94** 062505

[32] Hamamoto I 2014 *Phys. Rev.* C **89** 057301

[33] Podolyak Z 2012 *J. Phys. Conf. Series* **381** 012052

[34] Lucas R Jan/Feb 2001 *Europhysics News* p 5

[35] Oks E 2018 *Atoms* **6** 14

[36] Kryukov N and Oks E 2013 *Can. J. Phys.* **91** 715

[37] Ponomarev L I 1990 *Contemp. Phys.* **31** 219

[38] Nagamine K 2001 *Hyperfine Interact.* **138** 5

[39] Nagamine K and Ponomarev L I 2003 *Nucl. Phys.* A **721** C863

[40] Chelkowsky C, Bandrauk A D and Corkum P B 2004 *Laser Phys.* **14** 473

[41] Guffin J, Nixon G, Javorsek D II, Colafrancesco S and Fischbach E 2002 *Phys. Rev.* D **66** 123508

[42] Kryukov N and Oks E 2013 *Can. J. Phys.* **91** 715

[43] Kryukov N and Oks E 2014 *Can. J. Phys.* **92** 1405

[44] Rau A R P 1996 *J. Astrophys. Astr.* **17** 113

[45] Kryukov N and Oks E 2014 *Can. J. Phys.* **92** 1405

[46] Kryukov N and Oks E 2020 *Eur. Phys. J.* D **74** 82

[47] Kryukov N and Oks E 2019 *Int. Rev. Atom. Mol. Phys.* **9** 1

[48] Oks E 2020 *Can. J. Phys.* **98** 857

[49] Landau L D and Lifshitz E M 1965 *Quantum Mechanics* (Oxford: Pergamon)

[50] Bethe H A and Salpeter E E 1977 *Quantum Mechanics of One- and Two-Electron Atoms* (New York: Plenum Press)

[51] Oks E 2021 *Dynamics* **1** 1

[52] Kryukov N and Oks E 2021 *Open J. Microphys.* accepted

Chapter 2

Debye screening of circular Rydberg states of hydrogenic systems in collinear electric and magnetic fields of arbitrary strengths

A hydrogenic atom/ion of the nuclear charge Z under collinear electric (**F**) and magnetic (**B**) fields of arbitrary strengths was studied in paper [1]. The focus was on the circular Rydberg states of the electron, whose orbital plane was perpendicular to the magnetic field **B**.

The author of paper [1] presented analytical results on the classical ionization threshold F_c versus B, as well as on the energy at this threshold E_c versus B.

In paper [2], a brief overview of which we present here, the authors generalized the results from paper [1]. They considered a hydrogenic atom/ion under the same configuration of the electric and magnetic fields of arbitrary strengths, but the entire system being immersed into a plasma that provides the Debye screening. They demonstrated the reduction of the critical value of the electric field F_c for a variety of the magnetic field strengths. Below are some details.

Due to the Debye screening, the Coulomb potential should be substituted by its screened counterpart (atomic units $\hbar = e = m_e = 1$ were used in paper [2]):

$$\frac{Z}{r} \to \frac{Z}{r} e^{-r/a}. \tag{2.1}$$

In equation (2.1), a is the screening length. The classical Hamiltonian has the following form in the cylindrical coordinates:

$$H(\rho, z) = \frac{L^2}{2\rho^2} - \frac{Z}{\sqrt{\rho^2 + z^2}} e^{-\sqrt{\rho^2+z^2}/a} + Fz + \Omega L + \frac{\Omega^2 \rho^2}{2}. \tag{2.2}$$

In equation (2.2), $\Omega = B/(2c)$ is the Larmor frequency and L is the conserved z-component of the angular momentum (the z-axis being along the magnetic field **B**).

doi:10.1088/978-0-7503-3939-1ch2

The authors of paper [2] introduced the following scaled quantities:

$$v = \frac{Z}{L^2}\rho,\ w = \frac{Z}{L^2}z,\ f = \frac{L^4}{Z^3}F,\ \omega = \frac{L^3}{Z^2}\Omega,\ h = \frac{L^2}{Z^2}H,\ \varepsilon = \frac{L^2}{Z^2}E,\ \lambda = \frac{L^2}{Za}. \quad (2.3)$$

Then the scaled Hamiltonian h can be represented as follows:

$$h = \frac{1}{2v^2} - \frac{1}{\sqrt{w^2 + v^2}}e^{-\lambda\sqrt{w^2+v^2}} + fw + \omega + \frac{\omega^2 v^2}{2}. \quad (2.4)$$

By equating to zero the partial derivatives

$$\frac{\partial h}{\partial w} = 0,\ \frac{\partial h}{\partial v} = 0 \quad (2.5)$$

the authors of paper [2] obtained conditions for having the dynamical equilibrium. In particular, from the first equality in equation (2.5) they found

$$f = -\frac{w\left(1 + \lambda\sqrt{w^2 + v^2}\right)e^{-\lambda\sqrt{w^2+v^2}}}{(w^2 + v^2)^{3/2}}. \quad (2.6)$$

From equation (2.6) it is seen that for $f > 0$, it is required that $w < 0$.

From the second equality in equation (2.5), they got

$$\frac{\left(1 + \lambda\sqrt{w^2 + v^2}\right)e^{-\lambda\sqrt{w^2+v^2}}}{(w^2 + v^2)^{3/2}} + \omega^2 = \frac{1}{v^4}. \quad (2.7)$$

By combining equations (2.6) and (2.7), the authors of paper [2] derived the following relation:

$$\omega^2 - \frac{f}{w} = \frac{1}{v^4}. \quad (2.8)$$

It should be emphasized that equation (2.8 does not depend on λ).

From equation (2.8) it is easy to find the equilibrium value of the scaled coordinate $w(v, f, \omega)$:

$$w(v, f, \omega) = \frac{f}{\omega^2 - \frac{1}{v^4}}. \quad (2.9)$$

On substituting w from equation (2.9) into equation (2.6), the authors of paper [2] got the dependence of the scaled electric field f on v, ω, and λ implicitly:

$$e^{-\lambda k}(1 + \lambda k) = -(\omega^2 - \frac{1}{v^4})k^3 \quad (2.10)$$

where

$$k = \sqrt{v^2 + \left(\frac{f}{\omega^2 - \frac{1}{v^4}}\right)^2}. \tag{2.11}$$

For obtaining the dependence of the scaled electric field f on v, ω, and λ explicitely, the authors of paper [2] solved the above equations numerically. Then they substituted the obtained value value of f into the expression for w from equation (2.9) and got the dependence of w on v, ω, and λ. Thereafter they substituted the obtained values of f and w in equation (2.4).

As a result, the authors of paper [2] got the equilibrium value of the scaled energy $\varepsilon(v, \omega, \lambda)$. In this way, they found the parametric dependence of the scaled energy on the scaled electric field f (via the parameter v) for the fixed values of both the scaled magnetic field ω and the scaled screening parameter λ.

The authors of paper [2] plotted the dependence $\varepsilon(f)$ at some fixed values of the scaled magnetic field ω for the screening parameter $\lambda = 0.3$ in comparison with the case of $\lambda = 0$. They demonstrated that the screening diminishes the critical value of f required for ionization. This had to be expected because the screening brings the continuum lower.

Figure 2.1 shows this effect of the screening in its 'purest' form: it presents the situation where there is no magnetic field ($\omega = 0$) and thus eliminates the stabilizing role of the magnetic field.

The authors of paper [2] wrote:

'Physically, the presence of two branches in each of the plots (solid or dashed) within Figs. 1–3 has the following explanation. For each value of $f < f_c$, where f_c is

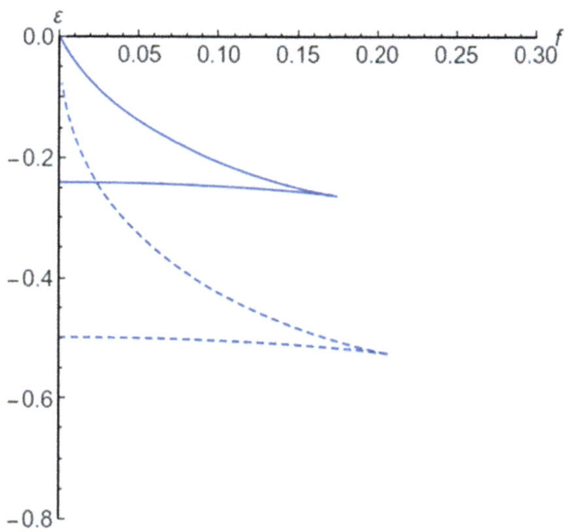

Figure 2.1. The plot of the parametric dependence $\varepsilon(f)$ at $\lambda = 0.3$ for $\omega = 0$ (solid curves) compared to the same plot at $\lambda = 0$ (dashed curves). [2] (2020). With permission of Springer.

the critical value beyond which there are no allowed values of the scaled energy ε, there are two values of the scaled equilibrium radius v of the orbit of the bound electron—as explained in paper [1] (see, especially figure 3.3 from paper [1]). One of the values correspond to the lower energy branch, another—to the upper energy branch. According to paper [1], the lower energy branch corresponds to the stable motion of the bound electron, while the upper energy branch corresponds to the unstable motion of the bound electron. This means that the critical scaled field f_c, where the two energy terms cross, is the classical ionization threshold: at $f = f_c$, the system can switch from the stable motion to the unstable motion and to 'slide' along the upper energy branch, so that the bound electron would end up in the continuum.'

For finding the dependence of the critical value of the electric field f_c on the scaled magnetic field ω the authors of paper [2] first used the equation $\partial\varepsilon/\partial v = 0$ to obtain the corresponding scaled radius of the orbit—that equation yielded:

$$\frac{4}{v^5}(k^2 - v^2) = \left(\omega^2 - \frac{1}{v^4}\right)\left(v - k\frac{dk}{dv}\Big|_c\right). \tag{2.12}$$

In equation (2.12), k is the solution of equation (10), while $(dk/dv)|_c$ was obtained by using the differentiation of equation (2.10) with respect to v:

$$\frac{dk}{dv}\bigg|_c = -\frac{4k^3}{v^5(\lambda e^{-\lambda k} - \lambda k(1 + \lambda k) + 3k^2(\omega^2 - \frac{1}{v^4}))}. \tag{2.13}$$

Then the authors of paper [2] substituted equation (2.13) in equation (2.12), while using the numerical solution of equation (2.10) with respect to k. Thereafter they obtained a numerical solution of the resulting equation with respect to v_c and found the critical value $v_c(\omega, \lambda)$, i.e., the value at the ionization threshold. Thereafter they substituted $v_c(\omega, \lambda)$ in $f(v, \omega, \lambda)$ and we got the dependence $f_c(\omega, \lambda)$, i.e., the critical

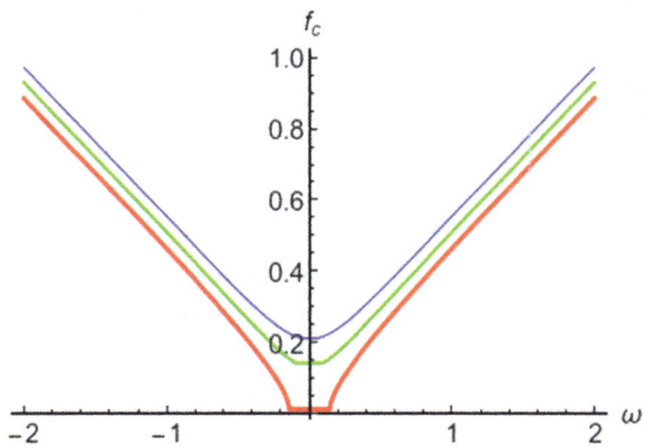

Figure 2.2. The plot of the dependence $f_c(\omega)$ at $\lambda = 0$ (blue, thin top curve), $\lambda = 0.4$ (green, thicker middle curve) and $\lambda = 0.6$ (red, thick bottom curve). [2] (2020). With permission from Springer.

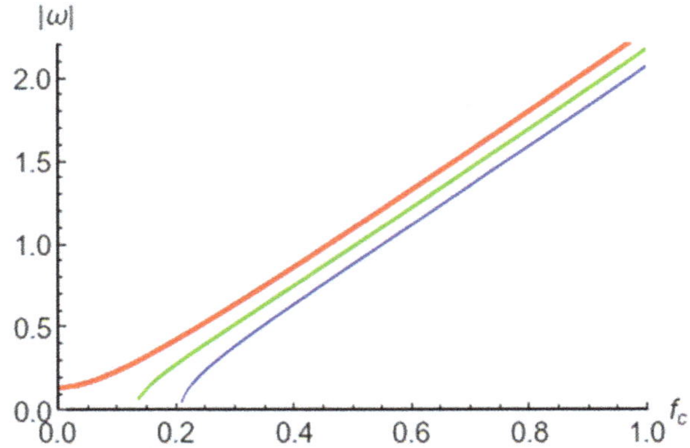

Figure 2.3. The plot of the dependence $|\omega|(f_c)$ at $\lambda = 0$ (blue, thin bottom curve), $\lambda = 0.4$ (green, thicker middle curve) and $\lambda = 0.6$ (red, thick top curve). [2] (2020). With permission from Springer.

electric field versus the scaled magnetic field ω for any fixed value of the screening parameter λ.

Figure 2.2 displays the critical electric field f_c versus the scaled magnetic field ω for three values of the screening parameter: $\lambda = 0$, 0.4 and 0.6. One can clearly see the effect of the screening: it diminished the critical electric field.

Figure 2.3 presents the dependence of the absolute value of the scaled magnetic field $|\omega|$ on the critical value of the scaled electric field f_c.

The critical electric field for various scaled magnetic fields and various scaled screening parameters can be obtained from the plot in figure 2.2 (and from the corresponding formulas above). Similarly, the scaled magnetic field for various scaled critical electric fields and for various scaled screening parameters can be obtained from the plot in figure 2.3 (and from the corresponding formulas above).

The authors of paper [2] wrote:

'For example, let us consider laser-produced plasmas emitting hydrogenic spectral lines that exhibit Langmuir-wave-caused dips (L-dips) in the profiles—such as, e.g., in experiments [3, 4]. The L-dips arise from the resonance coupling of the Langmuir waves with a quasistatic electric field in plasmas; the coupling is facilitated by the radiation hydrogenic atom or ion—see, e.g., papers [5–7]. The L-dip phenomenon allows, in particular, to accurately measure both the electron density N_e and the quasistatic electric field F involved in the resonance. There were dozens and dozens of experiments by various groups around the world at difference plasma machines, where such measurements were performed—see, e.g., books and reviews [8–11] and references therein.

Specifically, at laser-plasma experiments with the laser intensity $I >> 10^{18}$ W cm^{-2} (such as, e.g., the experiments [3, 4]), GigaGauss (GG) and even multi-GG magnetic fields are expected to be developed during relativistic laser–plasma interactions: these fields should be localized at the surface of the relativistic critical density—see, e.g., review [12] and references therein. The ultra-strong magnetic field expected to arise in such experiments (the experiments exhibiting the L-dips) can be estimated using the results of our paper as follows.

From the experiment one would know the nuclear charge of the radiating ion Z and the principal quantum number n of the last observable Ly line. From the L-dips one would determine the electron density N_e and the characteristic value of the quasistatic electric field F (responsible for the L-dips). This value of F can be substituted in the formulas from our paper as the critical electric field F_c. The temperature T can be estimated from the shape of the experimental spectral line profiles. Then one can calculate the screening parameter a (the Debye radius) and its scaled counterpart $\lambda \sim n^2/(Za)$. Next, one can estimate $f_c \sim F_c n^4/Z^3$. Then from our theoretical results for $\omega(f_c, \lambda)$, one can estimate ω. Next, one can estimate $\Omega(\text{a.u.}) \sim Z^2\omega(f_c, \lambda)/n^3$ and $\Omega(\text{s}^{-1}) = 4.13 \times 10^{16}\ \Omega(\text{a.u.})$, from where the estimate of the magnetic field can be obtained as $\text{B} = 2m_e c\ \Omega(\text{s}^{-1})/e$.'

References

[1] Oks E 2004 *Eur. Phys. J.* D **28** 171

[2] Kryukov N and Oks E 2020 *Eur. Phys. J.* D **74** 125

[3] Oks E *et al* 2017 *Opt. Express* **25** 1958

[4] Oks E *et al* 2017 *J. Phys. B: At. Mol. Opt. Phys.* **50** 245006

[5] Zhuzhunashvili A I and Oks E 1977 *Sov. Phys. JETP* **46** 1122

[6] Gavrilenko V P and Oks E A 1981 *Sov. Phys. JETP* **53** 1122

[7] Gavrilenko V P and Oks E A 1987 *Sov. J. Plasma Phys.* **13** 22

[8] Oks E 1995 *Plasma Spectroscopy: The Influence of Microwave and Laser Fields Springer Series on Atoms and Plasmas* vol 9 (Berlin: Springer)

[9] Oks E 2017 *Diagnostics of Laboratory and Astrophysical Plasmas Using Spectral Lineshapes of One-, Two-, and Three-Electron Systems* (Singapore: World Scientific)

[10] Dalimier E, Oks E and Renner O 2014 *Atoms* **2** 178

[11] Oks E, Dalimier E, Faenov A and Renner O 2014 *J. Phys. Conf. Series* **548** 012030

[12] Belyaev V S, Krainov V P, Lisitsa V S and Matafonov A P 2008 *Phys. – Uspekhi* **51** 793

Chapter 3

The shape of spectral lines of two-electron Rydberg atoms/ions: analytical solution

In this chapter we present a combination of two lines of research. One line is studies of helium atoms and helium-like ions. They are the simplest two-electron systems. They can be described theoretically with good accuracy, so that they can serve as calibrated probes for spectroscopic diagnostics of laboratory and astrophysical plasmas. The theoretical shapes of helium and helium-like spectral lines in plasmas, as well as their applications, can be found, e.g., in the following books [1–5] listed in chronological order.

Another line of research is investigations of how rotating electric fields affect atoms and ions. A number of authors studied how the rotating electric field affects a hydrogenic atom/ion. The purpose of those studies was to apply the results to the Stark broadening of hydrogenic spectral lines in plasmas.

In 1939–1940, Spitzer [6–8] obtained the wave functions of the energy level of the principal quantum number $n = 2$ of a hydrogenic ion while the rotating field was due to the passing ion. Spitzer approximated the actual hyperbolic trajectory of the passing ion by a straight line. In 1967, Ishimura [9] studied the effect of a uniformly rotating electric field of a constant magnitude on the splitting of the Ly-alpha line of hydrogen. He treated this setup as a model of the electric field of the passing charged particle. The authors of papers [6–9] obtained their analytical results by the direct solution of the corresponding Schrödinger equation for the first excited energy level. Therefore, their results could not be extended to the energy levels characterized by higher quantum numbers.

Hydrogen atoms have higher than geometrical symmetry: the O_4 symmetry discovered by Fock [10]. The manifestation of the O_4 symmetry of hydrogen atoms is an extra conserved vector quantity: the Runge–Lenz vector [11] (though it should be more properly called Hermann–Bernoulli–Laplace–Runge–Lenz vector [12].)

Thirty years after Spitzer's works [6–8], Lisitsa [13] used the O_4 symmetry of hydrogen atoms to bring up the possibility to produce the corresponding analytical

doi:10.1088/978-0-7503-3939-1ch3

solution for any principal quantum number n of a hydrogen atom—for the rotating electric field of Ishimura's model. Lisitsa demonstrated that this problem can be 'projected' on the problem of a hydrogen atom in crossed electric and magnetic fields. The latter problem has a known analytical solution given in a paper of 1969 by Demkov, Monozon, and Ostrovsky [14]. (For completeness it should be mentioned that Ishimura's model can correspond to one real situation, as noted by Lisitsa [15]: a hydrogen atom in a circularly-polarized laser field.)

Later, Lisitsa and Sholin [16] published the exact analytical solution for the Stark broadening of any hydrogen spectral line produced by one flight of a passing free electron (i.e., this was the binary case). Of course, they employed the O_4 symmetry of hydrogen atoms. Lisitsa and Sholin [16] produced explicit results for the shape of the Ly-αlpha line only. Later, their results for hydrogen atoms were extended by Greene, Cooper, and Smith [17] to spectral lines of hydrogenic ions—also under one flight of a passing electron (i.e., again the binary case).

Later Derevianko and Oks [18] obtained the exact analytical solution for the multi-particle (i.e., non-binary) case. Their general formalism they applied to the hydrogen Ly-alpha line in paper [18] and to the hydrogen Ly-beta line in paper [19].

In the present chapter we follow paper [20]. In that paper, the author analyzed the two-electron Rydberg atom/ion. Specifically, the system under consideration consisted of a nucleus of charge Z and two electrons: the inner electron denoted as 'e' and the outer (Rydberg) electron denoted as 'e*'. The average distance R_{av} of the outer electron e* from the nucleus was much greater than the average distance r_{av} of the inner electron e from the nucleus. The condition $r_{av} \ll R_{av}$ would be met if the principal quantum number n_2 of the outer electron would be just several times greater than the principal quantum number n_1 of the inner electron—because the distances r_{av} and R_{av} scale as n_1^2 and n_2^2, respectively.

The inequality $r_{av} \ll R_{av}$ ensures that the orbit of the outer electron is very close to the Kepler ellipse—specifically, the Kepler ellipse of an electron in a hydrogenic atom/ion of the effective nuclear charge $(Z-1)$, the electron revolving with the Kepler frequency Ω_2. Thus, the hydrogenic subsystem Z+e experiences the rotating electric field. The magnitude and the angular velocity of this rotating electric field depend on time.

One of the analytical results of paper [20] was the shape of any spectral line emitted by the subsystem Z+e in this case. The most detailed results were provided for the shape of the Ly-alpha line. It yielded a *peculiar result* as follows.

Since the subsystem Z+e is hydrogenic, one would expect it to exhibit a linear Stark effect with respect to the electric field of the outer electron. However, it turned out that the most intense components of the split Ly-alpha line exhibit a *quadratic* Stark effect with respect to the electric field of the outer electron. Below are the details.

This system emits two sets of spectral lines at significantly different frequencies. The spectral lines emitted by the outer electron are at the frequencies equal to $s\Omega_2$ (s = 1, 2. 3. …). Those are radiative transitions from the level n_2 to the level $(n_2 - s)$, where $s \ll n_2$. More accurate (than $s\Omega_2$) expressions for the frequencies of the spectral lines from this set can be found from the energy levels of the outer electron provided, e.g., by Nikitin and Ostrovsky [21].

The second set of the spectral lines is emitted by the inner electron at the frequencies

$$\omega = [Z^2(1 + 1/m)/2](1/n_0^2 - 1/n_1^2), \tag{3.1}$$

where m is the mass of the nucleus and n_0 is the principal quantum number of the lower level involved in the radiative transition. Here and below, atomic units are used.

This second set of the spectral lines was the focus of paper [20]. The corresponding frequencies could be in the optical range which makes them more easily observable—in distinction to the first set of spectral lines whose frequencies could be in the radio frequency range.

The Hamiltonian of the relative motion of the inner electron in the Z+e subsystem can be represented in the form

$$H(t) = p^2(1 + 1/m)/2 - Z/r - \mathbf{dF}(t), \quad \mathbf{d} = -(1 + Z/m)r/(1 + 1/m). \tag{3.2}$$

In equation (3.2), the quantity **dF** is the scalar product of the vector **F** of the electric field due to the outer electron and of the dipole moment vector **d**. (The notation **AB** throughout this chapter stands for the scalar product of any two vectors **A** and **B**.)

The Kepler frequency of the outer electron can be represented in the form

$$\Omega_2 = \left\{[(Z - 1)(1 + 2/m)]/[R_0^3(1 + 1/m)]\right\}^{1/2}. \tag{3.3}$$

In equation (3.3), the quantity R_0 is the major semi-axis of the elliptical orbit of the outer electron. The next step is to proceed to a rotating reference frame. Specifically, this is the frame rotating with the time-dependent angular velocity $\Omega_2(t)$ of the outer electron about the direction of the angular momentum **M**. After the transition to this rotating reference frame, the Hamiltonian of the relative motion of the inner electron from equation (3.2) acquires an additional term (see, e.g., book [22])

$$H(t) = p^2(1 + 1/m)/2 - Z/r + V(t), \quad V(t) = -\mathbf{dF}(t) - \mathbf{M\Omega}_2(t). \tag{3.4}$$

In equation (3.4), **M** and $\mathbf{M\Omega}_2$ are the angular momentum of the relative motion of the inner electron and its scalar product with the angular velocity vector $\mathbf{\Omega}_2$, respectively. The additional term $\mathbf{M\Omega}_2(t)$ in equation (3.4) is equivalent to some effective magnetic interaction. Therefore, equation (3.4) can be reformulated as follows

$$H(t) = p^2(1 + 1/m)/2 - Z/r - \mathbf{r}F_{\text{eff}}(t) - \boldsymbol{\mu}\mathbf{B}_{\text{eff}}, \tag{3.5}$$

where

$$\begin{aligned} \mathbf{B}_{\text{eff}}(t) &= -2c(1 + 1/m)\mathbf{\Omega}_2(t)/(1 - Z/m^2), \\ \boldsymbol{\mu} &= -(1 - Z/m^2)\mathbf{M}/[2c(1 + 1/m)]. \end{aligned} \tag{3.6}$$

In equations (3.5) and (3.6), $\mathbf{\mu}$ is the magnetic dipole moment. In equation (3.5), the vector $\mathbf{F}_{\rm eff}(t)$ stands for

$$\mathbf{F}_{\rm eff}(t) = -(1 + Z/m)\mathbf{F}(t)/(1 + 1/m),$$
$$\mathbf{F}(t) = [1 + \varepsilon \cos(\Omega_2 t + \beta)]^2 / [(1 - \varepsilon^2)^2 R_0^2]\mathbf{R}/R. \quad (3.7)$$

In equation (3.7), ε is the eccentricity of the Kepler ellipse of the outer electron and β is the initial phase.

The Hamiltonian in equations (3.4) or (3.5) is mathematically equivalent to the Hamiltonian of a hydrogenic atom/ion in crossed, time-dependent electric and magnetic fields. For the time-independent case, the classical analytical solution was known since 1925—see, e.g., Born's book [23]. The corresponding quantum analytical results were obtained by Demkov, Monozon, and Ostrovsky [14] in 1969. The classical and quantum solutions utilized the O_4 symmetry of hydrogenic systems and dealt with the Runge–Lenz vector [12]

$$\mathbf{A} = \mathbf{p}\text{x}\mathbf{M} - Z\mathbf{r}/r \text{ (classically)}, \quad \mathbf{A} = (\mathbf{p}\text{x}\mathbf{M} - \mathbf{M}\text{x}\mathbf{p})/2 - Z\mathbf{r}/r \text{ (quantally)}. \quad (3.8)$$

At this point it is important to bring the attention of the reader to Lisitsa–Sholin's paper [16]. In that paper the authors analyzed a hydrogen atom under one flight of a passing free electron. They employed the following fact concerning the time-dependent crossed electric and magnetic fields. For their problem, the ratio of the two fields remained time-independent despite both fields being dependent on time. This led to the existence of the two fixed 'directions of quantization'. Employing this fact, the authors of paper [16] showed that in the rotating reference frame, there is only a phase modulation of the atomic oscillator.

The author of paper [20] found a similar situation for the more complicated system under his consideration. Both the angular frequency $\mathbf{\Omega}_2(t)$ of the outer electron and its electric field $\mathbf{F}(t)$ at the location of the subsystem Z+e turned out to have the same time dependence, meaning that the ratio $F(t)/\Omega_2(t)$ was time-independent. This was because $\Omega_2(t) = (1 + 2/m)M/[(1 + 1/m)R^2(t)] = \text{const}/R^2(t)$ and $F(t) = 1/R^2(t)$, where $R(t)$ is the absolute value of the radius-vector of the outer electron.

Below we use the same notations as in paper [20]:

$$\mathbf{\Omega}_F(t) = (3/2)[r_0/(m_0 Z)]^{1/2} F_{\rm eff}(t), \quad \mathbf{N} = (m_0 r_0/Z)^{1/2}\mathbf{A}, \quad r_0 = n_1^2/(m_0 Z). \quad (3.9)$$

In equation (3.9), the quantity m_0 is the reduced mass of the inner electron:

$$m_0 = 1/(1 + 1/m). \quad (3.10)$$

From the physical point of view, the quantum expression for the quantity r_0 in equation (3.9) corresponds classically to the size of the major semi-axis of the elliptical orbit of the *inner* electron e. Also physically, the absolute value Ω_F of the vector $\mathbf{\Omega}_F$ (orthogonal to the vector $\mathbf{\Omega}_2$) is the instantaneous separation of the Stark sublevels in the field $F_{\rm eff}(t)$. From the classical point of view, the quantity Ω_F is the instantaneous frequency of the precession of the Kepler ellipse of the inner electron about the direction of the field $\mathbf{F}_{\text{eff}}$.

Next, similarly to papers [14, 16], the following four vectors have been introduced[1]:

$$\mathbf{J}_i(t) = [\mathbf{M} + (-1)^{i+1}\mathbf{N}(t)]/2, \quad \boldsymbol{\omega}_i(t) = \boldsymbol{\Omega}_2(t) + (-1)^{i+1}\boldsymbol{\Omega}_F(t); \quad i = 1, 2. \quad (3.11)$$

Then the interaction $V(t)$ in equation (3.4) can be reformulated as follows:

$$V(t) = -\mathbf{J}_1(t)\boldsymbol{\omega}_1(t) - \mathbf{J}_2(t)\boldsymbol{\omega}_2(t). \quad (3.12)$$

The directions of the vectors $\boldsymbol{\omega}_i(t)$ are time-independent because the ratio $q = \Omega_F(t)/\Omega_2(t) = \text{const}\ F(t)/\Omega_2(t)$ does not depend on time. This fact allows diagonalizing the Hamiltonian from equation (3.4) with the interaction $V(t)$ from equation (3.12) in terms of the instantaneous wave functions $u_{nn'n'}$, where the quantum number n' corresponds to a definite instantaneous projection of $\mathbf{J}_1$ on $\boldsymbol{\omega}_1$ and the quantum number n' corresponds to a definite instantaneous projection of $\mathbf{J}_2$ on $\boldsymbol{\omega}_2$—analogously to papers [14, 16].

The quantum numbers n' and n' take the following set of values: $-(n-1)/2$, $-(n-1)/2 + 1$, $-(n-1)/2 + 2$, …, $(n-1)/2$. Here and below for brevity, the subscript '1' at the quantity n has been omitted. The instantaneous 'energies' can be represented in the form

$$E_{nn'n_{''}}(t) = -Z^2/[2n^2(1 + 1/m)] - (n' + n')\omega(t). \quad (3.13)$$

The quantity $\omega(t)$ in equation (3.13) can be expressed as follows:

$$\omega(t) = |\ \omega_1(t)\ | = |\ \omega_2(t)\ | = [1 + q^2(n)]^{1/2}\Omega_2(t). \quad (3.14)$$

Here

$$\begin{aligned} q(n) &= \text{const}\ F(t)/\Omega_2(t) = [(1 + 1/m)r_0/Z]^{1/2}3(1 + Z/m)/[2(1 + 2/m)M] \\ &= 3n(1 + Z/m)(1 + 1/m)/[2(1 + 2/m)ZM]. \end{aligned} \quad (3.15)$$

By using the semiclassical expression $M = L + 1/2$, where L is the angular momentum quantum number of the outer electron, the quantity $q(n)$ can be reformulated as follows:

$$q(n) = 3n(1 + Z/m)(1 + 1/m)/[2(1 + 2/m)Z(L + 1/2)]. \quad (3.16)$$

It should be noted that

$$q(n) \sim n/[Z(L + 1/2)] \ll 1 \quad (3.17)$$

under the condition

$$L >> n/Z. \quad (3.18)$$

[1] The author of paper [20] used some notations from the classical solution presented in the book by Kotkin and Serbo, the 2nd Russian edition, problem 2.37 [24]—see also the 2nd English edition [25].

Under the inequality (3.18), the vectors $\boldsymbol{\omega}_i$ are directed at small angles ~ q compared to the direction of the vector $\boldsymbol{\Omega}_2$. Without the factor Z, the inequality (3.18) would be physically equivalent to the requirement of the non-penetration of the subsystem Z+e by the outer electron. However, the presence of the quantity Z in (3.18) relaxes this physical requirement. Also, for the semiclassical treatment of the outer electron it is required $L \gg 1$.

In the rotating frame, the instantaneous eigenfunctions are

$$\chi(t) = u_{nn'n''} \exp\left[-i\omega_0 t + \mathrm{i}(\mathrm{n}' + \mathrm{n}'')[1 + q^2(\mathrm{n})]^{1/2} \int_{t_0}^{t} \Omega_2(s)ds\right] \tag{3.19}$$

In equation (3.19), ω_0 is the unperturbed frequency of the spectral line resulting from the radiative transition of the inner electron between the upper level n_1 and the lower level n_0. The quantity $\Omega_2(t)$ is equal to $d\theta/dt$, where $\theta(t)$ is the variable describing the angular motion of the outer electron along its elliptical orbit. Therefore, equation (3.18) can be reformulated as follows:

$$\chi(t) = u_{nn'n''} \exp\left\{-i\omega_0 t + \mathrm{i}(n' + n'')[1 + q^2(n)]^{1/2}[\theta(t) - \theta(t_0)]\right\}. \tag{3.20}$$

Now we follow paper [20] to calculate the shape $S(\Delta\omega)$ of the corresponding spectral line. The shape $S(\Delta\omega)$, where $\Delta\omega$ is the frequency counted from the unperturbed frequency ω_0 of the spectral line, is the Fourier transform of the correlation function $\Phi(\tau)$ of the dipole moments of the subsystem Z+e (see, e.g., book [26]). For obtaining analytical results for $S(\Delta\omega)$ in a universal form, we follow the author of paper [20] to *measure* $\Delta\omega$ *in units of the Kepler frequency* Ω_2 *of the outer electron* (given by equation (3.3)) and *the time variables in units of* $1/\Omega_2$.

The lineshape can be presented in the following form

$$S(\Delta\omega) = \sum_{k=-1,0,1} \sum_{j} |\langle n_1 n_1' n_1'' \mid d_k \mid n_0 n_0' n_0'' \rangle|^2 J_{kj}. \tag{3.21}$$

In equation (3.21), j denotes a set of four quantum numbers (n_1', n_1', n_0', n_0') and d_k are the spherical components of the dipole moment defined according to Wigner [27] (in particular, $2^{1/2}d_{-1} = d_x + id_y = -2^{1/2}d_{+1}$). The quantity J_k in equation (3.21) has the following double integral expression, which includes, in particular, the averaging over the initial instant t_0:

$$J_{kj} = [1/(2\pi)^2] \int_{\text{period}} d\tau \exp(i\Delta\omega\tau) \int_{\text{period}} dt_0 \exp\left\{-i(\lambda_j + k)[\theta(\tau - t_0) - \theta(-t_0)]\right\}. \tag{3.22}$$

In equation (3.22), the integrals are to be calculated over any interval of the length 2π (which is the period of the revolution of the outer electron in units of $1/\Omega_2$) and

$$\lambda_j = (n_0' + n_0'')[1 + q^2(n_0)]^{1/2} - (n_1' + n_1'')[1 + q^2(n_1)]^{1/2}. \tag{3.23}$$

In equation (3.22) there is a factor k that takes values –1, 0, 1. This factor appears as a result of the transition from the instantaneous eigenfunctions $\chi(t)$ in the rotating frame (given by equation (3.20)) to the corresponding eigenfunctions in the laboratory frame. This transition is performed using Wigner's D-matrix $D^{(1)}(0, 0, -\theta(t))$—similarly to Lisitsa–Sholin's paper [16].

After the substitution $\tau_1 = \tau - t_0$, equation (3.22) takes the form:

$$J_{kj} = [1/(2\pi)^2] \int_{\text{period}} d\tau_1 \exp[i\Delta\omega\tau_1 - i(\lambda_j + k)\theta(\tau_1)] \int_{\text{period}} dt_0 \exp[i\Delta\omega t_0 + i(\lambda_j + k)\theta(-t_0)]. \tag{3.24}$$

By using the fact that $\theta(-t_0) = -\theta(t_0)$, equation (3.24) can be reduced to the square of a single integral:

$$J_{kj} = I_{kj}^2, \qquad I_{kj} = [1/(2\pi)] \int_{\text{period}} dt \exp\left[i\Delta\omega t - i(\lambda_j + k)\theta(t)\right]. \tag{3.25}$$

In paper [20], the integral I_k in equation (3.25) has been calculated using the change of the variable of the integration from t to the so-called eccentric anomaly η and employing the following well-known relations

$$t = (\eta - \varepsilon \sin \eta), \qquad \theta = 2 \tan^{-1}\left\{[(1 + \varepsilon)/(1 - \varepsilon)]^{1/2} \tan(\eta/2)\right\}. \tag{3.26}$$

In equation (3.26), ε is the eccentricity of the elliptical orbit of the outer electron. Finally, the integral I_{kj} takes the form

$$I_{kj} = [1/(2\pi)] \int_{-\pi}^{\pi} d\eta (1 - \varepsilon \cos \eta) \cos\Big[\Delta\omega(\eta - \varepsilon \sin \eta) - 2(\lambda_j + k) \tan^{-1}\{[(1 + \varepsilon)/(1 - \varepsilon)]^{1/2} \tan(\eta/2)\}\Big]. \tag{3.27}$$

It should be noted that the imaginary part of I_{kl} is actually zero despite its formal presence in equation (3.25).

Thus, there has been derived a universal analytical expression for the shape $S_{kj}(\Delta\omega)$ of the components of the spectral lines, emitted by the inner electron, in terms of the square $J_{kj} = I_{kj}^2$ of a one-fold integral $I_{kj}(\Delta\omega)$. This integral $I_{kj}(\Delta\omega)$ depends on the quantity λ_j given by equation (3.23) and on the eccentricity ε of the elliptical orbit of the outer electron. In its turn, the quantity λ_j depends on the combination of quantum numbers, which identify each component of the spectral line, and on the dimensionless parameters $q(n_1)$ and $q(n_0)$, where $q(n)$ is given by equation (3.15).

These general results have been illustrated in paper [20] by using as an example the Ly-alpha line ($n_1 = 2$, $n_0 = 1$). For the Ly-alpha line, equation (3.23) generates just the following three quantities

$$\lambda = 0, \pm[1 + q^2(2)]^{1/2}. \tag{3.28}$$

The relative intensities of the components of the Ly-alpha line are the same as from Lisitsa–Sholin's paper [16]. They are reproduced below in table 3.1. The sum of relative intensities of all components is normalized to unity. The components of the Ly-alpha line are labeled by their 'central frequency' $\lambda_j + k$, the latter quantity

Table 3.1. Central frequencies $\lambda_j + k$ and relative intensities of components of the Ly-alpha line emitted by the inner electron. The central frequency is the frequency at which the component has the primary maximum of the intensity for the zero eccentricity of the orbit of the outer, Rydberg electron. The frequencies are in units of the Kepler frequency Ω of the outer electron (given by equation (3.3)) and they are counted from the unperturbed frequency of the spectral line. The sum of relative intensities of all components is set to unity. Reprinted from [20].

Central frequency λ_j+k	$\pm[(1+q^2)^{1/2}-1]$	$\pm[(1+q^2)^{1/2}+1]$	±1	0
Relative intensity	$[(1+q^2)^{1/2}+1]/[10(1+q^2)]$	$[(1+q^2)^{1/2}-1]/[10(1+q^2)]$	$q^2/[5(1+q^2)]$	1/5

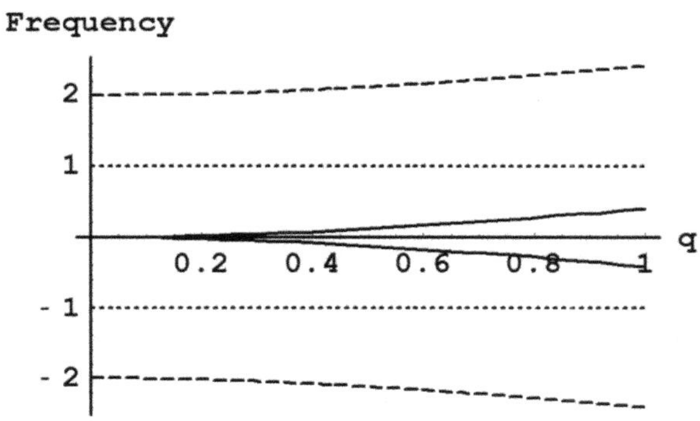

Figure 3.1. Central frequencies of the Ly-alpha line components versus the parameter q (given by equations (3.15) and (3.16)). Components at the central frequencies $\pm[(1+q^2)^{1/2}-1]$, $\pm[(1+q^2)^{1/2}+1]$, and $\pm$ 1 are shown by solid, dashed, and dotted lines, respectively. The frequency of the 0-component coincides with the abscissa axis. The central frequency is defined as the frequency, at which the component has the primary maximum of the intensity for the zero eccentricity of the orbit of the outer, Rydberg electron. The frequencies are in units of the Kepler frequency Ω_2 of the outer electron (given by equation (3.3)) and they are counted from the unperturbed frequency of the spectral line. Reprinted from [20].

significantly controling the integral in equation (3.27). For each line component, the location of the primary maximum of its intensity for $\varepsilon = 0$ serves as the central frequency. A reminder: the frequencies are counted from the unperturbed frequency of the Ly-alpha line and are measured in units of the Kepler frequency Ω_2 of the outer electron.

Figure 3.1 presents the central frequencies of the spectral line components (in units of Ω_2) versus the parameter q in the following way. Components at the central frequencies $\pm[(1+q^2)^{1/2}-1]$, $\pm[(1+q^2)^{1/2}+1]$, and $\pm$ 1 are shown by solid, dashed, and dotted lines, respectively. Obviously, the frequency of the 0-component coincides with the abscissa axis.

Figure 3.2 presents the dependence of the relative intensities of the Ly-alpha line components on the parameter q in the following way. Intensities of the components having the central frequencies $\pm[(1+q^2)^{1/2}-1]$, $\pm[(1+q^2)^{1/2}+1]$, $\pm$ 1, and 0 are shown by solid, dashed, dotted, and dash-dotted lines, respectively.

Figure 3.2 demonstrates that at $q^2 << 1$, about 80% the entire intensity of the spectral line is concentrated in the components characterized by the central

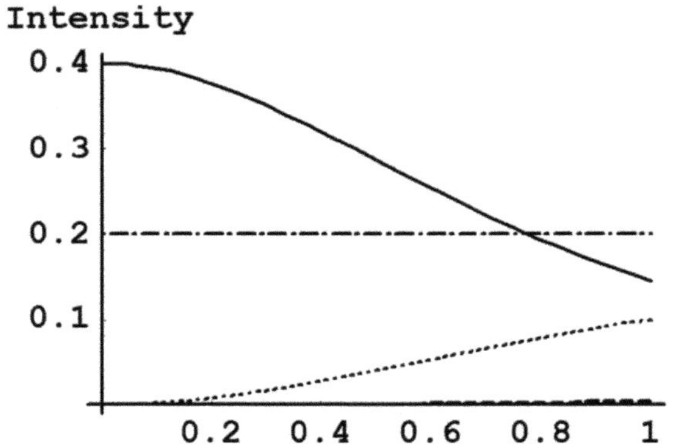

Figure 3.2. Relative intensities of the Ly-alpha line components versus the parameter q (given by equations (3.15) and (3.16)). Intensities of the components having the central frequencies $\pm[(1+q^2)^{1/2}-1]$, $\pm[(1+q^2)^{1/2}+1]$, $\pm$ 1, and 0 are shown by solid, dashed, dotted, and dash-dotted lines, respectively. The sum of relative intensities of all components is set to unity. Reprinted from [20].

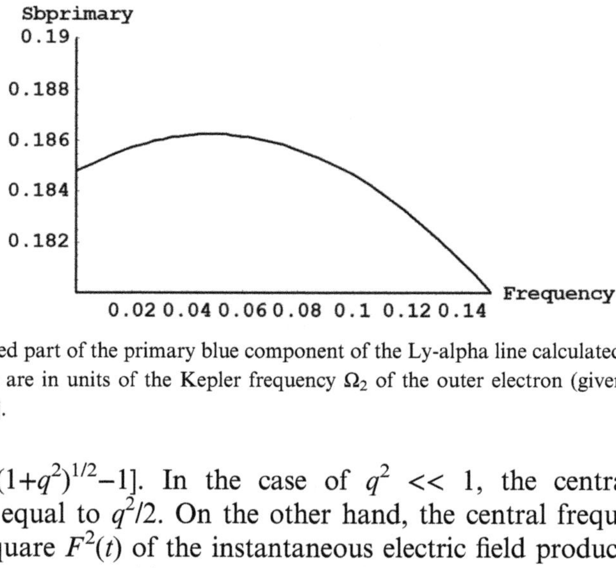

Figure 3.3. Magnified part of the primary blue component of the Ly-alpha line calculated at $\varepsilon = 0$ and $q^2 = 1/10$. The frequencies are in units of the Kepler frequency Ω_2 of the outer electron (given by equation (3.3)). Reprinted from [20].

frequencies $\pm[(1+q^2)^{1/2}-1]$. In the case of $q^2 << 1$, the central frequency is approximately equal to $q^2/2$. On the other hand, the central frequency is proportional to the square $F^2(t)$ of the instantaneous electric field produced by the outer electron—because $q = \Omega_F(t)/\Omega_2(t) = \text{const}\ F(t)/\Omega_2(t)$. This means that at $q^2 << 1$, the the electric field caused by the outer electron leads to the *quadratic* Stark effect for the most intense components of the line. (We note that the case of $q^2 << 1$ is the actual situation as long as $L^2 >> (n/Z)^2$ according to equation (3.18).)

This is a *counter-intuitive* result. Since the subsystem Z+e is hydrogenic, one would think that the Stark effect in an external electric field should have been linear.

Figure 3.3 illustrates this small quadratic shift of the maximum of the profile of the component labeled as $[(1+q^2)^{1/2}-1]$ (this is the primary component in the blue part of the Ly-alpha line). The line profile, whose magnified part (Sbprimary) close to the unperturbed frequency is shown in figure 3.3, was calculated at $\varepsilon = 0$ and $q^2 = 1/10$.

The author of paper [20] provided the following physical explanation of this counter-intuitive result. In the case of $L \gg n/Z$, the primary effect is the rotation of the electric field caused by the outer electron. Therefore, the dominant effect in the splitting of the spectral line is owing to the *amplitude modulation* of the $(Z+e)$-oscillator by the rotating electric field. (We recall that the amplitude modulation is the change in the amplitude of the emitted electromagnetic wave.) The amplitude modulation leads to the quadratic Stark effect.

The shape of the entire Ly-alpha line can be represented in the form:

$$\begin{aligned} J(\Delta\omega, \varepsilon) = {} & \left\{[(1+q^2)^{1/2}+1]/[10(1+q^2)]\right\}\left\{ I^2[\Delta\omega, \varepsilon, (1+q^2)^{1/2}-1] + I^2[\Delta\omega, \varepsilon, 1-(1+q^2)^{1/2}]\right\} \\ & \left\{[(1+q^2)^{1/2}-1]/[10(1+q^2)]\right\}\left\{I^2[\Delta\omega, \varepsilon, (1+q^2)^{1/2}+1] + I^2[\Delta\omega, \varepsilon, -(1+q^2)^{1/2}-1]\right\} + \\ & \left\{q^2/[5(1+q^2)]\right\}\left\{I^2[\Delta\omega, \varepsilon, 1] + I^2[\Delta\omega, \varepsilon, -1]\right\} + (1/5)\delta(\Delta\omega). \end{aligned} \quad (3.29)$$

In the last term in equation (3.29), $\delta(\Delta\omega)$ is the Dirac δ-function. It conveys the fact that the rotating electric field caused by the outer electron does not broaden the 'unshifted' component of the Ly-alpha line. Of course, this 'unshifted' component has a relatively narrow Lorentzian profile corresponding to the natural width, which is the inverse of the natural lifetime of the $n = 2$ state of the Z+e subsystem.

Figure 3.4 shows, as an example, the total profile (Stot) of all 'shifted' components at $\varepsilon = 0$ and two values of q^2: $q^2 = 1/5$ (solid line) and $q^2 = 1/25$ (dashed line). Figure 3.4 demonstrates that at $q^2 = 1/5$, the profile exhibits a pedestal (or 'shoulders') at $|\Delta\omega| = (1{-}1.4)$, while at $q^2 = 1/25$ secondary maxima are more pronounced.

Figure 3.5 presents, as an example, the total profile (Stot) of all 'shifted' components at $q^2 = 1/5$ and two values of ε: $\varepsilon = 0.9$ (solid line) and $\varepsilon = 0$ (dashed

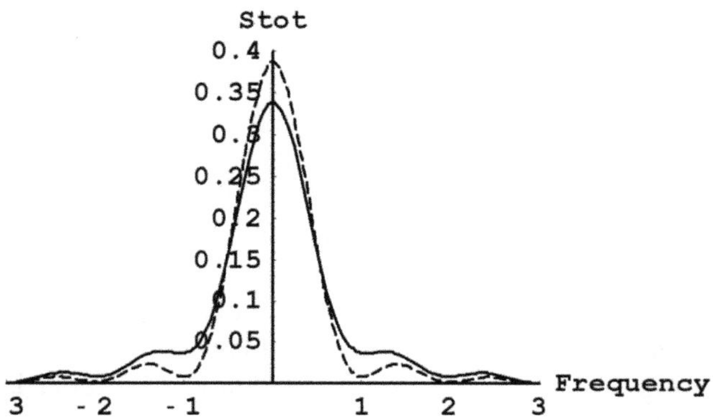

Figure 3.4. Total profile of all 'shifted' components at $\varepsilon = 0$ and two values of q^2: $q^2 = 1/5$ (solid line) and $q^2 = 1/25$ (dashed line). The frequencies are in units of the Kepler frequency Ω_2 of the outer electron (given by equation (3.3)). Reprinted from [20].

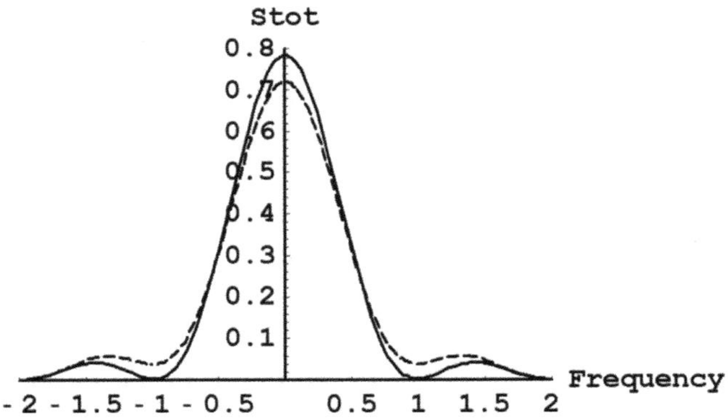

Figure 3.5. Total profile of all 'shifted' components at $q^2 = 1/5$ and two values of ε: $\varepsilon = 0.9$ (solid line) and $\varepsilon = 0$ (dashed line). The frequencies are in units of the Kepler frequency Ω_2 of the outer electron (given by equation (3.3)). Reprinted from [20].

line). Figure 3.5 shows that a relatively large eccentricity ($\varepsilon = 0.9$) of the orbit of the outer electron makes the secondary maxima more distinct compared to the case of the circular orbit.

The integrals $I_{jk}[\Delta\omega, \varepsilon, \lambda_j(q) + k]$ entering the shapes of the individual components of the Ly-alpha line are functions of two continuous parameters (in addition to depending on the argument $\Delta\omega$): ε and $\lambda_j(q) + k$. Therefore, for making plots of the function $I_{jk}^2[\Delta\omega, \varepsilon, \lambda_j(q) + k]$ at some few representative values of ε, it would be necessary (for each chosen ε) to produce a three-dimensional plot of I^2 versus both $\Delta\omega$ and q.

However, the components labeled by $\lambda_j + k = \pm 1$ are the exceptions. Their profile $J_0(\Delta\omega, \varepsilon)$ does not depend on q:

$$J_0(\Delta\omega, \varepsilon) = [1/(2\pi)^2]\left\{\int_{-\pi}^{\pi} d\eta(1 - \varepsilon\cos\eta)\cos[\Delta\omega(\eta - \varepsilon\sin\eta) \pm 2\tan^{-1}\{[(1+\varepsilon)/(1-\varepsilon)]^{1/2}\tan(\eta/2)\}]\right\}^2. \tag{3.30}$$

In equation (3.30), the plus sign corresponds to $\lambda_j + k = -1$ and the minus sign corresponds to $\lambda_j + k = 1$. Figure 3.6 presents the shape (S) of the component corresponding to $\lambda_j + k = 1$ for the eccentricity (of the orbit of the outer electron) $\varepsilon = 0.8$ (solid line), $\varepsilon = 0.4$ (dashed line), and $\varepsilon = 0.8$ (dotted line).

Figure 3.6 demonstrates that at $\varepsilon = 0$, the profile of this component is symmetric with respect to $\Delta\omega = 1$ ($\Delta\omega = 1$ is the central frequency for this component). However, the profile shows a significant asymmetry (with respect to $\Delta\omega = 1$) for non-zero values of the eccentricity ε. (We note that for the Ly-alpha component corresponding to $\lambda_j + k = -1$, the profiles in figure 3.6 should be reflected with respect to the vertical axis.)

By using the smallness of the quantity q (i.e., $q^2 \ll 1$), simplified analytical formulas can be obtained for other components of the Ly-alpha line. These results,

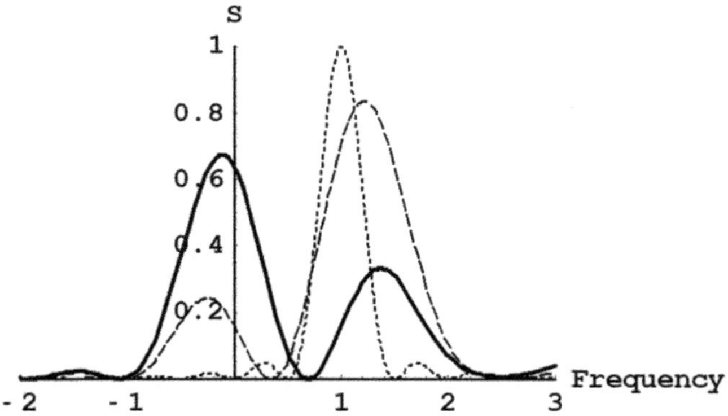

Figure 3.6. The shape S of the Ly-alpha component corresponding to $\lambda_j + k = 1$ for the following values of the eccentricity ε of the elliptical orbit of the outer electron: $\varepsilon = 0.8$ (solid line), $\varepsilon = 0.4$ (dashed line), $\varepsilon = 0$ (dotted line). The frequency is in units of the Kepler frequency Ω_2 of the outer electron (given by equation (3.3)). Reprinted from [20].

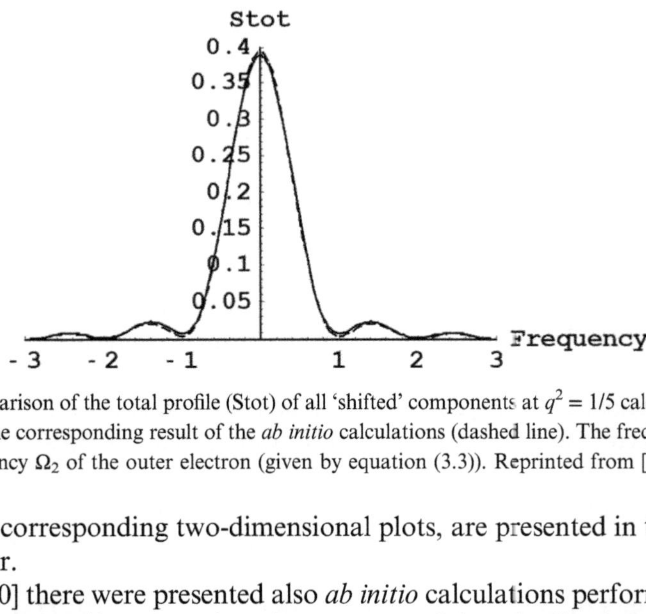

Figure 3.7. Comparison of the total profile (Stot) of all 'shifted' components at $q^2 = 1/5$ calculated analytically (solid line) with the corresponding result of the *ab initio* calculations (dashed line). The frequency is in units of the Kepler frequency Ω_2 of the outer electron (given by equation (3.3)). Reprinted from [20].

as well as the corresponding two-dimensional plots, are presented in the Addendum to this chapter.

In paper [20] there were presented also *ab initio* calculations performed by solving numerically the Hamilton equations for a system of three bodies 1, 2, 3, where body 1 was the outer electron, body 2—the nucleus, and body 3—the inner electron. These numerical calculations were performed along the lines of paper [28]. The spatial orientations and momenta of the electrons were chosen randomly in accordance with the Monte Carlo technique.

Figure 3.7 presents the comparison of the total profile (Stot) of all 'shifted' components at $q^2 = 1/5$ calculated in two different ways. The results of the analytical calculations are shown by the solid line while the corresponding results of the *ab initio* calculations are shown by the dashed line. It is seen that there is practically no difference between the results of these two calculations.

In paper [20] there was also presented a comparison of the above dynamical Stark broadening by the outer electron (for brevity, DSBOE) with other possible broadening mechanisms and/or models. The comparison was presented both for laboratory atomic experiments without any plasma involved and for some astrophysical and laboratory plasmas. As an example, there was used the Ly-alpha line of the ionized helium (He II 30.38 nm) emitted by the Z+e subsystem under the DSBEO. The principal quantum number of the outer electron was $n_2 = 10$.

The width caused by the DSBOE $\Delta\omega_{DSBRE}$ is equal approximately to the Keppler frequency of the Rydberg electron:

$$\Delta\omega_{DSBOE} \sim \omega_{au}/n^3 = 4.13 \times 10^{16}\text{s}^{-1}/10^3 = 4.13 \times 10^{13}\text{s}^{-1}. \quad (3.31)$$

In equation (3.31), ω_{au} is the atomic unit of frequency.

The DSBOE $\Delta\omega_{DSBOE}$ exceeds the natural width $\Delta\omega_{nat} = 1.25 \times 10^9$ s^{-1} by four orders of magnitude.

For purely atomic experiments without involving any plasma, the Doppler width at the room temperature, i.e., at $T = 0.026$ eV, is $\Delta\omega_D = 0.958 \times 10^{11}$ s^{-1}.

For the conditions of He II emitting regions of solar flares (chromosphere and transition region plasmas), where $T \sim 2$ eV, the Doppler width is $\Delta\omega_D = 8.42 \times 10^{11}$ s^{-1}.

For the conditions of edge plasmas of magnetic fusion devices, where $T \sim 4$ eV, the Doppler width is $\Delta\omega_D = 1.19 \times 10^{12}$ s^{-1}.

It is seen that in the above cases, the Doppler broadening is smaller than the DSBOE width $\Delta\omega_{DSBOE}$ by one or two orders of magnitude.

In paper [20] the comparison was also done with the dynamical Stark broadening by plasma electrons (DSBE) and with the dynamical Stark broadening by plasma ions (DSBI) for the following types of plasmas.

For the conditions of He II emitting regions of solar flares, where the electron density is ~ 10^{13} cm^{-3} and $T = 2$ eV, the DSBE width is $\Delta\omega_{DSBE} = 1.2 \times 10^8$ s^{-1} and the DSBI is $\Delta\omega_{DSBI} = 1.9 \times 10^9$ s^{-1}.

For the conditions of the edge plasmas of magnetic fusion devices, where the electron density is ~ 10^{14} cm^{-3} and $T = 4$ eV, the DSBE width is $\Delta\omega_{DSBE} = 4.6 \times 10^8$ s^{-1} and the DSBI is $\Delta\omega_{DSBI} = 1.3 \times 10^{10}$ s^{-1}.

It is seen that in the above cases of astrophysical and laboratory plasmas, the DSBE and the DSBI are smaller than the DSBOE by three or four or five orders of magnitude.

It is worth underlining that that the DSBRE and the DSBE/DSBI are *different kinds of the Stark broadening*, and that the DSBRE was never included in any calculations of the DSBE/DSBI.

The new type of spectral line profiles, calculated analytically in paper [20], is significantly different from the line profiles caused by the dynamical Stark broadening in plasmas. This could result in a variety of line shapes: from the simplest Lorentzian line profile to much more complicated line profiles (due to both the phase modulation, and the amplitude modulation, and the nonadiabatic effects)—see, e.g., the exact analytical results from papers [16–19] and the review in book [29].

Physically, the main difference between the spectral line broadening calculated in paper [20] and the DSBE/DSBI is the following. The spectral line broadening calculated in paper [20] is due to the periodic or quasi-periodic motion of the Rydberg electron, while in the DSBE/DSBI the motion of perturbing electrons is aperiodic. This qualitative distinction leads to the significant difference in spectral line shapes.

Addendum. Asymptotic results for the shapes of the Ly-alpha components at $q^2 << 1$

The shape of the two line components, which are the most intense at $q^2 << 1$(the components labeled by $\lambda_j + k = \pm[(1 + q^2)^{1/2} - 1]$ corresponding to the solid line in figure 3.2) is the following

$$J_{-}\left\{\Delta\omega, \varepsilon, \pm[(1 + q^2)^{1/2} - 1]\right\} = J_1(\Delta\omega, \varepsilon) \pm q^2 J_2(\Delta\omega, \varepsilon), \qquad (3.A.1)$$

where the plus sign corresponds to $\lambda_j + k = (1 + q^2)^{1/2} - 1$ and the minus sign corresponds to $\lambda_j + k = 1 - (1 + q^2)^{1/2}$. In equation (3.A.1) the functions $J_1(\Delta\omega, \varepsilon)$ and $J_2(\Delta\omega, \varepsilon)$ are as follows:

$$\begin{aligned} J_1(\Delta\omega, \varepsilon) &= [1/(2\pi)^2]\left\{\int_{-\pi}^{\pi} d\eta(1 - \varepsilon\cos\eta)\cos[\Delta\omega(\eta - \varepsilon\sin\eta)]\right\}^2 \\ &= [\sin(\pi\Delta\omega)/(\pi\Delta\omega)]^2 \end{aligned} \qquad (3.A.2)$$

and

$$\begin{aligned} J_2(\Delta\omega, \varepsilon) = [\sin(\pi\Delta\omega)/(\pi^2\Delta\omega)]\int_{-\pi}^{\pi} d\eta(1 - \varepsilon\cos\eta) \\ \sin[\Delta\omega(\eta - \varepsilon\sin\eta)]\tan^{-1}\{[(1 + \varepsilon)/(1 - \varepsilon)]^{1/2}\tan(\eta/2)\}. \end{aligned} \qquad (3.A.3)$$

From equation (3.A.2) it is seen that the main part $J_1(\Delta\omega, \varepsilon)$ of the profile of these two most intense components has a simple analytical expression—the expression not depending on the eccentricity ε of the elliptical orbit of the Rydberg electron. Figure 3.A.1 shows the corresponding shape (S).

Figure 3.A.2 demonstrates the shape (S) of the correction term $J_2(\Delta\omega, \varepsilon)$ in equation (3.A.1) for the component labeled by $\lambda_j + k = (1 + q^2)^{1/2} - 1$ for various eccentricities of the elliptical orbit of the Rydberg electron: $\varepsilon = 0.8$ (solid line), $\varepsilon = 0.4$ (dashed line), and $\varepsilon = 0.8$ (dotted line). For $\varepsilon = 0$, the correction term $J_2(\Delta\omega, 0)$ has the following analytical expression.

$$J_2(\Delta\omega, 0) = [1 - \cos(2\pi\Delta\omega) - \pi\Delta\omega\sin(2\pi\Delta\omega)]/(2\pi^2\Delta\omega^3). \qquad (3.A.4)$$

For the spectral line component labeled by $\lambda_j + k = 1 - (1 + q^2)^{1/2}$, the profiles demonstrated in figure 3.A.2 should be reflected with respect to the vertical axis.

It should be noted that in the sum of the profiles of these two components of the Ly-alpha line, there is no correction term proportional to q^2: it vanishes. Indeed, from equation (3.A.1) it follows that $J_{-}[\Delta\omega, \varepsilon, (1 + q^2)^{1/2} - 1] + J_{-}[\Delta\omega, \varepsilon, 1 - (1 + q^2)^{1/2}] = 2J_1(\Delta\omega, \varepsilon)$, where $J_1(\Delta\omega, \varepsilon)$ is presented in equation (3.A.2) and is illustrated in figure 3.A.1. A nonvanishing correction to the sum of the shapes of these two components is ~ q^4.

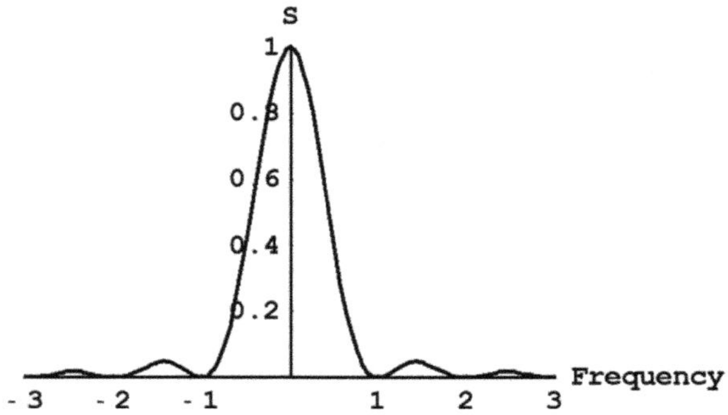

Figure 3.A.1. The primary part of the shape S of *each* of the two components of the Ly-alpha line that are the most intense at $q^2 << 1$. These are components labeled by $\lambda_j + k = \pm[(1 + q^2)^{1/2} - 1]$. The frequency is in units of the Kepler frequency Ω_2 of the outer electron (given by equation (3.3)). Reprinted from [20].

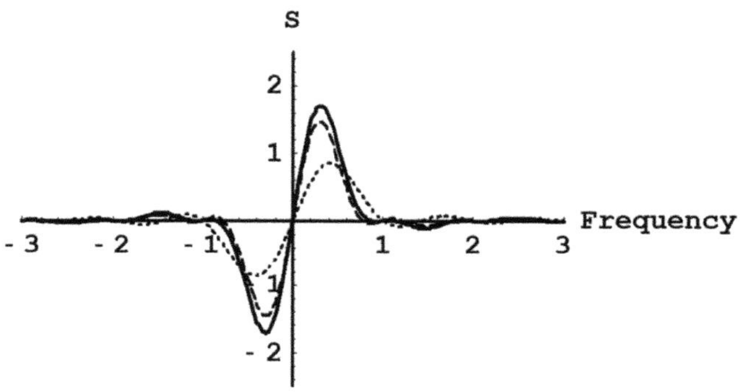

Figure 3.A.2. The shape S of the correction term $J_2(\Delta\omega, \varepsilon)$ in equation (3.A.1) for the component labeled by $\lambda_j + k = (1 + q^2)^{1/2} - 1$ for the following values of the eccentricity ε of the orbit of the outer electron: $\varepsilon = 0.8$ (solid line), $\varepsilon = 0.4$ (dashed line), $\varepsilon = 0$ (dotted line). The frequency is in units of the Kepler frequency Ω_2 of the outer electron (given by equation (3.3)). Reprinted from [20].

The other two shifted components are labeled by $\lambda_j + k = \pm[(1 + q^2)^{1/2} + 1]$. They correspond to the dashed line in figure 3.2. Their profiles can be presented as follows

$$J_+\left\{\left[\Delta\omega,\ \varepsilon,\ \pm(1 + q^2)^{1/2} + 1\right]\right\} = J_3(\Delta\omega, \varepsilon) \pm q^2 J_4(\Delta\omega, \varepsilon), \tag{3.A.5}$$

where the plus sign corresponds to $\lambda_j + k = (1 + q^2)^{1/2} + 1$ and the minus sign corresponds to $\lambda_j + k = -(1 + q^2)^{1/2} - 1$. In equation (3.A.5), the profile $J_3(\Delta\omega, \varepsilon)$ is

$$_3(\Delta\omega, \varepsilon) = [1/(2\pi)^2]$$

$$\left\{\int_{-\pi}^{\pi} d\eta (1 - \varepsilon \cos\eta) \cos\left[\Delta\omega(\eta - \varepsilon \sin\eta) \pm 4\tan^{-1}\left\{[1 + \varepsilon]/[1 - \varepsilon]^{1/2} \tan(\eta/2)\right\}\right]\right\}^2 \tag{3.A.6}$$

and $J_4(\Delta\omega, \varepsilon)$ is represented by the following product of two integrals

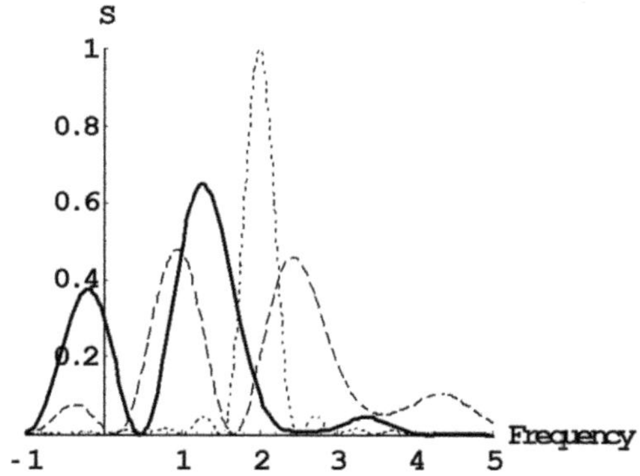

Figure 3.A.3. The primary part of the shape S of the component of the Ly-alpha labeled by $\lambda_j + k = (1 + q^2)^{1/2} + 1$ for $q^2 << 1$ for the following values of the eccentricity ε of the orbit of the outer electron: $\varepsilon = 0.8$ (solid line), $\varepsilon = 0.4$ (dashed line), $\varepsilon = 0$ (dotted line). The frequency is in units of the Kepler frequency Ω_2 of the outer electron (given by equation (3.3)). Reprinted from [20].

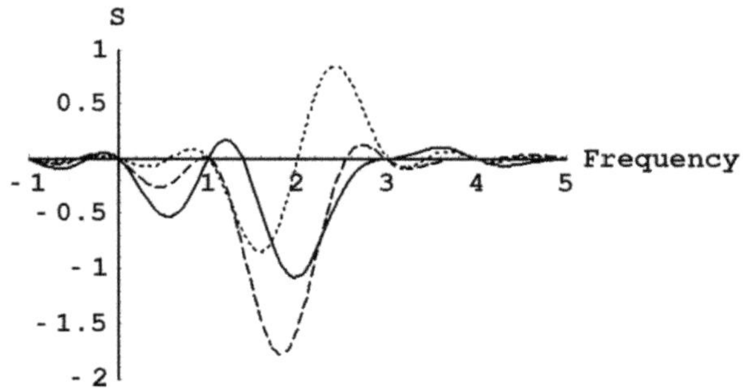

Figure 3.A.4. The shape S of the correction term $J_4(\Delta\omega, \varepsilon)$ in equation (3.A.5) for the component labeled by $\lambda_j + k = (1 + q^2)^{1/2} + 1$ for the following values of the eccentricity ε of the orbit of the outer electron: $\varepsilon = 0.8$ (solid line), $\varepsilon = 0.4$ (dashed line), $\varepsilon = 0$ (dotted line). The frequency is in units of the Kepler frequency Ω_2 of the outer electron (given by equation (3.3)). Reprinted from [20].

$$J_4(\Delta\omega, \varepsilon) = [1/(2\pi)^2]\left\{\int_{-\pi}^{\pi} d\eta(1 - \varepsilon\cos\eta)\right.$$
$$\left.\cos[\Delta\omega(\eta - \varepsilon\sin\eta) \pm 4\tan^{-1}\{[(1+\varepsilon)/(1-\varepsilon)]^{1/2}\tan(\eta/2)\}]\right\} \tag{3.A.7}$$
$$\left\{\int_{-\pi}^{\pi} d\eta(1 - \varepsilon\cos\eta)\sin[\Delta\omega(\eta - \varepsilon\sin\eta) \pm 4\tan^{-1}\{[(1+\varepsilon)/(1-\varepsilon)]^{1/2}\}]\right.$$
$$\left.\tan^{-1}\{[(1+\varepsilon)/(1-\varepsilon)]^{1/2}\tan(\eta/2)\}\right\}.$$

The shapes (S) of the profiles $J_3(\Delta\omega, \varepsilon)$ and $J_4(\Delta\omega, \varepsilon)$ for the component labeled by $\lambda_j + k = (1 + q^2)^{1/2} + 1$ are displayed in figures 3.A.3 and 3.A.4, respectively, for $\varepsilon = 0.8$ (solid line), $\varepsilon = 0.4$ (dashed line), and $\varepsilon = 0$ (dotted line). The corresponding profiles for the component labeled by $\lambda_j + k = -(1 + q^2)^{1/2} - 1$ can be obtained from the profiles in figures 3.A.3 and 3.A.4 by the reflection with respect to the vertical axis.

References

[1] Oks E 1995 *Plasma Spectroscopy: The Influence of Microwave and Laser Fields Springer Series on Atoms and Plasmas* vol 9 (New York: Springer)
[2] Griem H R 1997 *Principles of Plasma Spectroscopy* (Cambridge: Cambridge University Press)
[3] Fujimoto T 2004 *Plasma Spectroscopy* (Oxford: Clarendon)
[4] Kunze H-J 2009 *Introduction to Plasma Spectroscopy* (Berlin: Springer)
[5] Oks E 2017 *Diagnostics of Laboratory and Astrophysical Plasmas Using Spectral Lines of One-, Two-, and Three-Electron Systems* (Hackensack, NJ: World Scientific)
[6] Spitzer L 1939 *Phys. Rev.* **55** 699
[7] Spitzer L 1939 *Phys. Rev.* **56** 39
[8] Spitzer L 1940 *Phys. Rev.* **58** 348
[9] Ishimura T 1967 *J. Phys. Soc. Japan* **23** 422
[10] Fock V 1935 *Z. Physik* **98** 145
[11] Hughes J W B 1967 *Proc. Phys. Soc. (London)* **91** 810
[12] Goldstein H 1976 *Am. J. Phys.* **44** 1123
[13] Lisitsa V S 1971 *Optics Spectrosc.* **31** 468
[14] Yu N, Demkov B S, Monozon and Ostrovsky V N 1969 *Zh. Eksp. Teor. Fiz.* **57** 1431
[15] Lisitsa V S 1977 *Sov. Phys. Uspekhi* **20** 603
[16] Lisita V S and Sholin G V 1972 *Sov. Phys. JETP* **34** 484
[17] Greene R L, Cooper J and Smith E W 1975 *J. Quant, Spectrosc. Rad. Transfer* **15** 1025
[18] Derevianko A and Oks E 1996 *Physics of Strongly Coupled Plasmas* ed W D Kraeft and M Schlanges (Singapore: World Scientific) p 286 and p 291
[19] Derevianko A and Oks E 1999 *Proc. 14th Int. Conf. Spectral Line Shapes, AIP Conf. Proc.* **467** 148
[20] Oks E 2017 *J. Phys. B: At. Mol. Opt. Phys.* **50** 115001
[21] Nikitin S I and Ostrovsky V N 1978 *J. Phys. B: At. Mol. Phys.* **11** 1681
[22] Landau L D and Lifshitz E M 1960 *Mechanics* (Oxford: Pergamon) sections 39, 40
[23] Born M 1925 *Vorlesungen Über Atommechanik* (Berlin: Springer) ; 1927 *The Mechanics of the Atom* (London: Bell and Sons)
[24] Kotkin G L and Serbo V G 1977 *Sbornik Zadach po Klassicheskoj Mechanike (Collection of Problems in Classical Mechanics)* 2nd edn (Moscow: Nauka)
[25] Kotkin G L and Serbo V G 2020 *Exploring Classical Mechanics: A Collection of 350+ Solved Problems for Students, Lecturers, and Researchers* 2nd Revised and Enlarged English Edn (Oxford: Oxford University Press)
[26] Sobel'man I I 1972 *An Introduction to the Theory of Atomic Spectra* (Oxford: Pergamon)
[27] Wigner E P 1959 *Group Theory and its Application to the Quantum Mechanics of Atomic Spectra* (New York: Academic)
[28] Gavrilenko V P and Oks E 2007 *Phys. Scripta* **76** 43
[29] Oks E 2006 *Stark Broadening of Hydrogen and Hydrogenlike Spectral Lines in Plasmas: The Physical Insight* (Oxford: Alpha Science Int.)
[30] Oks E and Spectrosc J Q 2015 *Rad. Transfer* **152** 74
[31] Iglesias C A 2010 *High Energy Density Phys.* **6** 318

Chapter 4

Helium atoms and helium-like ions in the high-frequency laser field

Analytical results on hydrogenic atoms or ions in a high-frequency laser field have been presented in papers [1–3]. The Authors of papers [2, 3] concentrated on the classical description of Rydberg states. The laser frequency ω was considered to significantly exceed the Kepler frequency $\omega_K = m_e e^4/(n^3\hbar^3)$ of the highly-excited hydrogen atom: $\omega \gg \omega_K$. (In the previous sentence, $n \gg 1$ is the principal quantum number; m_e and e are the electron mass and charge, respectively.) Therefore, the Rydberg atom was treated as the slow subsystem and the laser field was treated as the fast subsystem. In paper [2], there was utilized the generalized Kapitza's method of effective potentials [4, 5].

In the situations where the laser field is linearly-polarized or circularly-polarized, there is a rich physics behind this fundamental problem, as shown in papers [1–3]. Namely, the symmetry of the system is higher than the geometrical one. Indeed, geometrically the system is characterized by the axial symmetry leading to the conservation of the angular momentum projection M_z on the axis of the symmetry. However, it turned out that the square of the angular momentum M^2 was also conserved.

Besides, interesting celestial analogies were pointed out. In the situation where the laser field is linearly-polarized, the system is mathematically equivalent to a satellite revolving around an oblate planet. The latter celestial system is known to demonstrate two precessions simultaneously: the precession of the plane of the satellite orbit and the precession within the plane of the satellite orbit. In the situation where the laser field is circularly-polarized, the system is mathematically equivalent to a satellite revolving around a prolate planet (a fictitious planet) displaying both types of the precession.

In the present chapter we follow paper [6] where the authors studied a Rydberg helium atom or a Rydberg helium-like ion (specifically, one electron is highly-excited) subjected to a high-frequency laser field. The generalized effective potential method [2, 7] (see also book [8]) was employed.

doi:10.1088/978-0-7503-3939-1ch4

The authors of paper [6] concentrated on circular Rydberg states. There are many papers presented theoretical and experimental studies of circular Rydberg states (e.g., papers [9–36]). The importance of studying circular Rydberg states is warranted by the following. First, for so-called coherent states described by quantum mechanics, circular Rydberg states are their classical analogs. Second, circular Rydberg states exhibit long lifetimes (with respect to the spontaneous emission) and anisotropic cross-sections of collisions, so that circular Rydberg states are utilized in experiments on inhibited spontaneous emission and in experiments on cold Rydberg atoms [37, 38]. Third, in the analytical method of quantum mechanics based on the expansion in terms of the small parameter $1/n$, classical circular Rydberg states serve as the primary term of the expansion (see, e.g., paper [39]).

The authors of paper [6] demonstrated that the orbital plane of the Rydberg electron undergoes a precession and provided the analytical result for the precession frequency. The latter was found to be different from the case of a hydrogen atom or a hydrogen-like ion. Because of the precession of the orbit of the Rydberg electron, the system should emit not only the spectral line at the Kepler frequency, but also satellites separated from the main line by multiples of the precession frequency.

In paper [6] there was also found another effect of the high-frequency laser field on the He atom or He-like ion. It was a red shift of the Rydberg electron energy. It was analytically calculated and its dependence on parameters of the system was analyzed. Below are the details.

The He atom or the He-like ion is considered to have the outer Rydberg electron and the inner electron in state 1s. The combination of the nucleus of charge Z and the spherically-symmetric charge distribution due to the inner electron creates the following electric potential (see, e.g., paper [40]):

$$\Phi(r) = \frac{Z-1}{r} + \left(Z\mu + \frac{1}{r}\right)e^{-2Z\mu r}. \tag{4.1}$$

In equation (4.1), $\mu = M_n m_e/(M_n + m_e)$ is the reduced mass of the quasi-nucleus (i.e., of 'nucleus—inner electron' pair), where m_e is the electron mass and M_n is the mass of the nucleus); r is the separation of the outer electron from the nucleus. The authors of paper [6] utilized atomic units in their paper: $\hbar = e = m_e = 1$. The He atom or the He-like ion is under a high-frequency laser field, the amplitude of the field being F and frequency of the field being ω.

First, the authors of paper [6] analyzed the situation where the laser field is linearly polarized. For the outer electron, the corresponding Hamiltonian in the spherical polar coordinates is

$$H = H_0 + zF\cos\omega t,\ H_0 = \frac{1}{2\mu_1}\left(p_r^2 + \frac{p_\theta^2}{r^2} + \frac{p_\varphi^2}{r^2\sin^2\theta}\right) - \Phi(r). \tag{4.2}$$

In equation (4.2), the z-axis is chosen along the laser field $\mathbf{F}$; $\mu_1 = m_e(M_n + m_e)/(M_n + 2m_e)$ is the reduced mass of the 'quasi-nucleus—outer electron' pair.

After the application of the formalism of effective potentials [2, 4, 5, 7, 8], a time-independent term is added to the Hamiltonian H_0. The 0-th order effective potential is

$$U_0 = \frac{1}{4\omega^2}[V, [V, H_0]] = \frac{F^2}{4\mu_1\omega^2} \tag{4.3}$$

where $V = zF$. In equation (4.3), $[P, Q]$ are the Poisson brackets. The additional term U_0 represents a coordinate-independent energy shift. Therefore, it does not change the dynamics of the system.

The next (first order) term in the effective potential does affect the dynamics of the system—the term having the following form

$$\begin{aligned} U_1 &= \frac{1}{4\omega^4}[[V, H_0], [[V, H_0], H_0]] \\ &= -\frac{F^2(Z\mu)^3}{8\mu_1^2\omega^4 x^3}((Z - 1)(1 + 3\cos 2\theta) \\ &+ ((1 + 2x)(1 + 2x^2) + (3 + 6x + 6x^2 + 4x^3)\cos 2\theta)e^{-2x}) \end{aligned} \tag{4.4}$$

where

$$x = \mu Z r. \tag{4.5}$$

Thus, the effective potential energy becomes:

$$U_{\text{eff}} = U + U_0 + U_1. \tag{4.6}$$

In equation (4.6), U is the unperturbed potential energy.

For a highly excited hydrogen atom in a linearly-polarized high frequency laser field, the effective potential energy (obtained in paper [2]) was

$$U_{\text{eff}}(r, \theta) = -\frac{e^2}{r} - \frac{\gamma}{r^3}(3\cos^2\theta - 1),\ \gamma = \frac{e^4F^2}{4m_e^2\omega^4}. \tag{4.7}$$

It is mathematically equivalent to the potential energy of a satellite orbiting an oblate planet (see, e.g., book [40]). So, the unperturbed elliptical orbit (whether the satellite orbit or electron orbit) is engaged simultaneously in the precession of the orbital plane about the axis of symmetry and into the precession in the plane of the orbit. The corresponding precession frequencies differ from each other, though they are of the same order of magnitude.

In the situation considered by the authors of paper [6], the effective potential energy turned out to be more complicated. Therefore, for obtaining explicit analytical results they proceeded to the case of circular orbits of the outer electron. In this case the only one remaining dynamic variable is θ (since r = const). In this situation, the effective potential energy reduces 'almost' to the form (4.7), but it differs from (4.7) by an additional θ-independent form, as presented below.

Equations (4.4) can be rewritten as follows

$$U_1 = -\frac{F^2(Z\mu)^3(Z-1)}{4\mu_1^2\omega^4x^3}\left((3\cos^2\theta - 1)\left(1 + g(x)\frac{e^{-2x}}{3(Z-1)}\right) + \frac{3f(x) - g(x)}{2}\frac{e^{-2x}}{3(Z-1)}\right) \tag{4.8}$$

where

$$f(x) = (1 + 2x)(1 + 2x^2),\ g(x) = 3 + 6x + 6x^2 + 4x^3. \tag{4.9}$$

Then the total energy becomes

$$U_{\text{eff}} \approx -\frac{Z-1}{r} + \Delta U_1(r) + \Delta U(r, \cos\theta). \tag{4.10}$$

In equation (4.10), the second term is now the constant shift of the energy, while the third term causes the precession.

From equations (4.8) and (4.7), it is seen that the term

$$g(x)\frac{e^{-2x}}{3(Z-1)} \tag{4.11}$$

constitutes a relative correction to the precession frequency, while the term

$$\frac{3f(x) - g(x)}{2}\frac{e^{-2x}}{3(Z-1)} = 4x^3\frac{e^{-2x}}{3(Z-1)} \tag{4.12}$$

represents the shift of the energy. After allowing for the factor in front of the right side of (4.8), the shift of the energy becomes

$$\delta E = -\frac{F^2(Z\mu)^3}{3\mu_1^2\omega^4}e^{-2x}. \tag{4.13}$$

Taking into account that for circular orbits, the outer electron energy is $E = -(Z-1)/(2r)$, and utilizing equation (4.5), the shift of the energy can be represented in the form

$$\delta E = -\frac{F^2(Z\mu)^3}{3\mu_1^2\omega^4}e^{\frac{\mu Z(Z-1)}{E}}. \tag{4.14}$$

Figure 4.1 displays the shift of the energy versus the unperturbed electron energy for $Z = 4$, $F = 1$, $\omega = 100$. For these parameters, the Kepler frequency of the outer electron

$$\omega_K = \frac{1}{Z-1}\sqrt{\frac{8|E|^3}{\mu_1}} \tag{4.15}$$

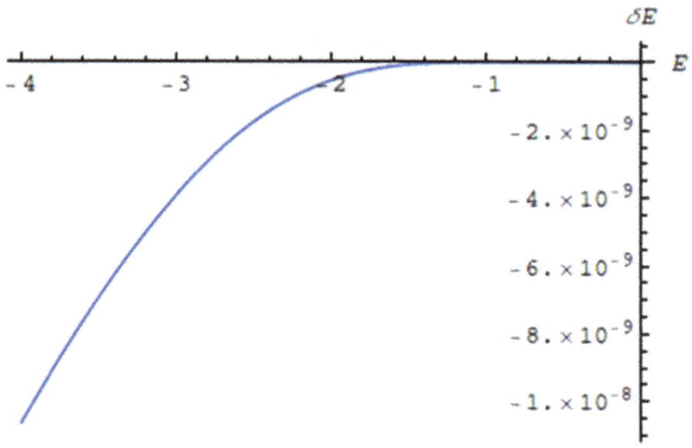

Figure 4.1. The energy shift versus the unperturbed electron energy for $Z = 4$, $F = 1$, $\omega = 100$. Reprinted from [6] with permission from De Gruyter (CC BY 4.0).

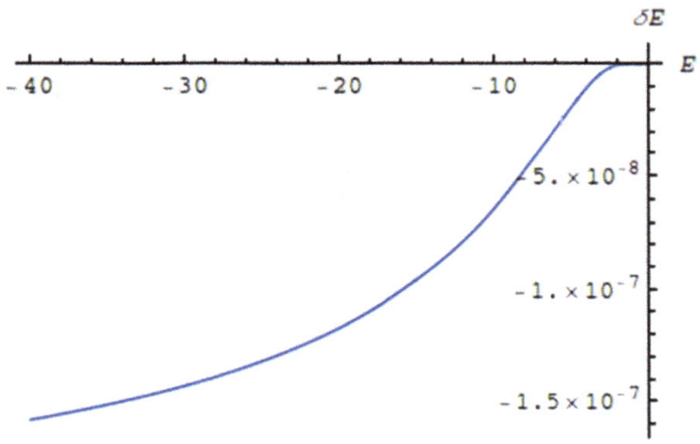

Figure 4.2. The energy shift versus the unperturbed electron energy E for $Z = 4$, $F = 1$, $\omega = 100$ for large negative values of E. Reprinted from [6] with permission from De Gruyter (CC BY 4.0).

is much smaller than the laser frequency, as required for the validity of the method of effective potentials.

With the increase of the absolute value of the unperturbed energy of the outer electron, the absolute value of the energy shift increases from zero to the limit of—$F^2(Z\mu)^3/(3{\mu_1}^2\omega^4)$—see figure 4.2.

Figure 4.3 shows the shift of the energy versus the nuclear charge Z at $F = 1$ and $\omega = 100$ for two values of the unperturbed energy: $E = -2$ (blue, solid line) and $E = -5$ (red, dashed line). It is seen that the dependence on the nuclear charge is non-monotonic. This is a *counter-intuitive result*. In the approximation $\mu = \mu_1 = 1$, the minimum is located at $Z_m \approx (1 + (1 + 24|E|)^{1/2})/4$.

Figure 4.4 presents the shift of the energy versus both the unperturbed energy of the outer electron and the nuclear charge for $F = 1$ and $\omega = 100$. Figure 4.4 shows that with the increase of the unperturbed energy of the outer electron, the minimum of the shift of the energy shift indeed moves to higher values of Z—in accordance with the above formula for Z_m.

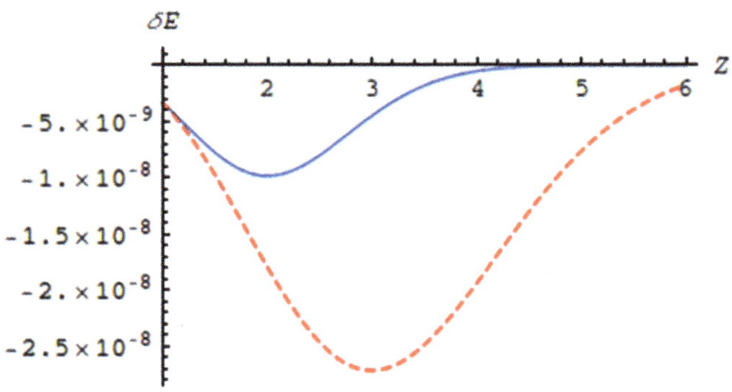

Figure 4.3. The energy shift versus the nuclear charge Z for $E = -2$ (blue, solid line) and $E = -5$ (red, dashed line) for $F = 1$, $\omega = 100$. Reprinted from [6] with permission from De Gruyter (CC BY 4.0).

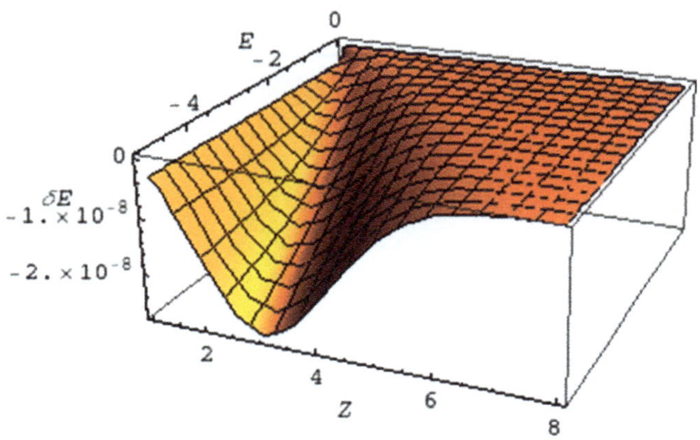

Figure 4.4. The dependence of the energy shift on Z and E for $F = 1$, $\omega = 100$. Reprinted from [6] with permission from De Gruyter (CC BY 4.0).

The effective potential energy of the outer electron can be rewritten in the form:

$$U_{\text{eff}} = -\frac{(Z-1)e^2}{r} - \frac{(Z-1)\gamma}{r^3}(3\cos^2\theta - 1). \tag{4.16}$$

The effective potential energy for a satellite orbiting an oblate planet is (according, e.g., to book [41]):

$$V_E = -\frac{GM_E m}{r} - \frac{GM_E m \mid I_2 \mid R^2}{2r^3}(3\cos^2\theta - 1). \tag{4.17}$$

In equation (4.17), I_2 is a constant characterizing the relative difference between the polar and equatorial diameters of the Earth, R is the equatorial radius of the Earth, M_E and m are the Earth and the satellite masses, respectively. It is seen that the effective potentials from equations (4.16) and (4.17) are mathematically equivalent.

According to book [41], the frequency Ω of the precession of the satellite orbit is (in units of the Kepler frequency):

$$\frac{\Omega}{\omega_{E,K}} = \frac{3 \mid I_2 \mid}{2}\left(\frac{R}{p}\right)^2 \cos i \tag{4.18}$$

where

$$\omega_{E,K} = \sqrt{\frac{G(M_E + m)}{A_s^3}} \tag{4.19}$$

is the Kepler frequency of the satellite. In equation (4.18), p is the semi-latus rectum of the unperturbed elliptical orbit and i is the *inclination* (i.e., the angle between the equatorial plane of the Earth the plane of the satellite orbit. In equation (4.19), A_s and m are the major semi-axis of the satellite unperturbed elliptical orbit and the satellite mass, respectively.

After redefining some quantites in the effective potential energy of the satellite from equation (17), namely

$$GM_E m = (Z - 1)e^2, \mid I_2 \mid R^2 = \frac{2\gamma}{e^2}. \tag{4.20}$$

Equation (4.17) would become identical to equation (4.16). On substituting $|I_2|R^2 = 2\gamma/e^2$ in equation (4.18), one gets the corresponding ratio of frequencies for the outer electron in the He atom or He-like ion:

$$\left.\frac{\Omega}{\omega_K}\right|_0 = \frac{3\gamma}{e^2 p^2}\cos i. \tag{4.21}$$

The Kepler frequency ω_K entering equation (4.21) was specified by equation (4.15); the subscript '0' refers to the unperturbed case.

For circular orbits the semi-latus rectum has the following form:

$$p = r = \frac{Z - 1}{2 \mid E \mid}. \tag{4.22}$$

Then equation (4.21) becomes

$$\left.\frac{\Omega}{\omega_K}\right|_0 = \frac{12\gamma|E|^2}{(Z - 1)^2}\cos i \tag{4.23}$$

In equation (4.23), the quantity γ has the following expression (according to equation (4.8)):

$$\gamma = \frac{F^2}{4\mu_1^2\omega^4}. \tag{4.24}$$

For circular orbits, the quantity x entering equation (4.23) has the following dependence on the unperturbed energy:

$$x(E) = \mu Z r = \frac{\mu Z(Z - 1)}{2 \mid E \mid}. \tag{4.25}$$

We recall that equations (4.21)–(4.25) related to the unperturbed circular orbit. For determining the dynamics of the perturbed 'circular orbit' it is sufficient to drop the second term in the effective potential from equation (4.10) (this term is constant) and to deal with the following truncated effective potential $U_{\text{eff, tr}}$:

$$U_{\text{eff,tr}} = -\frac{Z-1}{r} - \frac{(Z-1)\gamma}{r^3}\left(1 + g(x)\frac{e^{-2x}}{3(Z-1)}\right)(3\cos^2\theta - 1). \tag{4.26}$$

In equation (4.25), $g(x)$, r, and $x(E)$ are defined by equations (4.9), (4.22), and (4.25), respectively. So, in units of the Kepler frequency of the outer electron, the frequency of the precession of its orbit is the following:

$$\frac{\Omega}{\omega_K} = \frac{3\gamma}{r^2}\left(1 + g(x)\frac{e^{-2x}}{3(Z-1)}\right)\cos i = \frac{12\gamma\,|E|^2}{(Z-1)^2}\left(1 + g(x)\frac{e^{-2x}}{3(Z-1)}\right)\cos i. \tag{4.27}$$

Thus, the relative correction to the precession frequency can be represented in the form

$$\frac{\Delta\Omega}{\Omega} = g(x(E))\frac{e^{-2x(E)}}{3(Z-1)}. \tag{4.28}$$

Figure 4.5 shows the relative correction to the precession frequency of the orbital plane of the outer electron versus the electron energy for several values of the nuclear charge values of Z.

It is seen that at large absolute values of the energy of the Rydberg electron, the correction $\Delta\Omega/\Omega$ tends to the limit of $1/(Z-1)$.

Next, the authors of paper [6] analyzed the situation where the laser field is circularly polarized:

$$F = F(\mathbf{e}_x \cos\omega t + \mathbf{e}_y \sin\omega t) \tag{4.29}$$

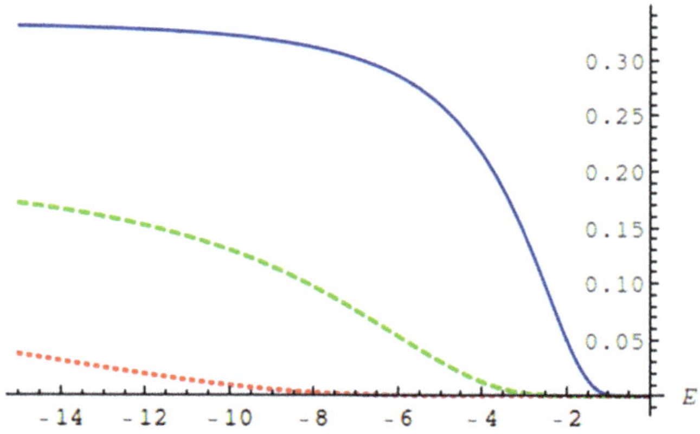

Figure 4.5. Dependence of the relative correction to the precession frequency of the orbital plane of the Rydberg electron on the electron energy for $Z = 4$ (blue solid curve), $Z = 6$ (green dashed curve), and $Z = 9$ (red dotted curve). Reprinted from [6] with permission from De Gruyter (CC BY 4.0).

Here $\mathbf{e_x}$ and $\mathbf{e_y}$ are the unit vectors along the x- and y-axes, the z-axis being chosen perpendicular to the polarization plane. The Hamiltonian for the outer electron is

$$H = H_0 + xF\cos\omega t + yF\sin\omega t \tag{4.30}$$

where H_0 is defined in equation (4.2). After introducing the notation

$$V = xF = Fr\,\sin\theta\,\cos\varphi,\ W = yF = Fr\,\sin\theta\,\sin\varphi \tag{4.31}$$

and engaging the formalism of effective potentials [2, 7, 8], the authors of paper [6] derived the following zeroth-order effective potential

$$U_0 = \frac{1}{4\omega^2}([V, [V, H_0]] + [W, [W, H_0]]) = \frac{F^2}{2\mu_1\omega^2} \tag{4.32}$$

and the corresponding first-order effective potential

$$\begin{aligned} U_1 = {} & \frac{1}{4\omega^4}([[V, H_0], [[V, H_0], H_0]] + [[W, H_0], [[W, H_0], H_0]]) \\ & + \frac{-1}{2\omega^3}[[V, H_0], [W, H_0]] \\ = {} & \frac{F^2(Z\mu)^3}{8\mu_1^2\omega^4x^3}((Z - 1)(1 + 3\cos 2\theta) \\ & + ((1 + 2x + 2x^2 - 4x^3) + (3 + 6x + 6x^2 + 4x^3)\cos 2\theta)e^{-2x}). \end{aligned} \tag{4.33}$$

The quantity x being defined by (4.5). Equation (4.33) can be rewritten in the form analogous to equation (4.8):

$$\begin{aligned} U_1 = {} & \frac{F^2(Z\mu)^3(Z - 1)}{4\mu_1^2\omega^4x^3}\left((3\cos^2\theta - 1)\left(1 + g(x)\frac{e^{-2x}}{3(Z - 1)}\right)\right. \\ & \left. + \frac{3f_1(x) - g(x)}{2}\frac{e^{-2x}}{3(Z - 1)}\right). \end{aligned} \tag{4.34}$$

It differs from (4.8) in two ways: by the sign and by a different function $f_1(x)$:

$$f_1(x) = 1 + 2x + 2x^2 - 4x^3. \tag{4.35}$$

From equation (4.34) it is easy to see that the shift of the energy now is

$$\delta E = -\frac{2F^2(Z\mu)^3}{3\mu_1^2\omega^4}e^{-2x}. \tag{4.36}$$

This shift is two times greater than in the case of the linear polarization. As for the relative correction to the precession frequency of the orbital plane, it turns out to be exactly equal to the corresponding result for the linear polarization case.

In summary, the authors of paper [6] presented an analytical study of a He atom or He-like ion, where one electron is in the Rydberg state, under a high-frequency laser field, They employed the generalized method of the effective potentials [2, 7, 8].

The focus of the study was on circular Rydberg states. The authors of paper [6] found that the orbital plane of the outer electron precesses and calculated the precession frequency, which turned out to be different from the corresponding precession frequency for a hydrogen atom or hydrogen-like ion. They pointed out that the observable manifestation of this precession would be the appearance of satellites (in the radiation spectrum) separated from the main spectral line by multiples of the precession frequency.

The second effect found by the authors of paper [6] was that the energy of the outer electron had an additional contribution: a red shift, which they calculated analytically. They demonstrated a monotonic increase of the absolute value of the red shift with the growth of the unperturbed binding energy of the outer electron. They also showed that the red shift depends on the nuclear charge non-monotonically: as the nuclear charge increases, the absolute value of the red shift first grows, but then reaches a maximum, and subsequently diminishes. This is a *counter-intuitive result.*

References

[1] Gavrilenko V P, Oks E and Radchik A V 1985 *Opt. Spectrosc.* **59** 411

[2] Nadezhdin B B and Oks E 1986 *Sov. Tech. Phys. Lett.* **12** 512

[3] Oks E, Davis J E and Uzer T 2000 *J. Phys. B: At. Mol. Opt. Phys.* **33** 207

[4] Kapitza P L 1951 *Sov. Phys. JETP* **21** 588

[5] Kapitza P L 1951 *Uspekhi Fiz. Nauk* **44** 7

[6] Kryukov N and Oks E 2021 *Open Phys.* **19** 11

[7] Nadezhdin B B 1986 Radiatsionnye i relativistskie effekty v atomakh i ionakh *Radiative and Relativistic Effects in Atoms and Ions* (Moscow: Scientific Council of the USSR Academy of Sciences on Spectroscopy) p 222 (in Russian)

[8] Oks E 2020 *Analytical Advances in Quantum and Celestial Mechanics: Separating Rapid and Slow Subsystems* (Bristol: IOP Publishing) appendix A

[9] Kryukov N and Oks E 2020 *Eur. Phys. J.* D **74** 125

[10] Kryukov N and Oks E 2020 *Eur. Phys. J.* D **74** 82

[11] Oks E 2020 *Can. J. Phys.* **98** 857

[12] Kryukov N and Oks E 2016 *J. Phys. B: At. Mol. Opt. Phys.* **49** 035002

[13] Kryukov N and Oks E 2014 *Eur. Phys. J.* D **68** 171

[14] Kryukov N and Oks E 2014 *Can. J. Phys.* **92** 1405

[15] Kryukov N and Oks E 2013 *J. Phys. B: Atom. Mol. Opt. Phys.* **46** 245701

[16] Kryukov N and Oks E 2013 *Can. J. Phys.* **91** 715

[17] Kryukov N and Oks E 2012 *Can. J. Phys.* **90** 647

[18] Tan J N, Brewer S M and Guise N D 2011 *Phys. Scr.* **T144** 014009

[19] Dehesa J S, Lopez-Rosa S, Martinez-Finkelshtein A and Yanez R J 2010 *Int. J. Quantum Chem.* **110**

[20] Nandi T 2009 *J. Phys. B: At. Mol. Opt. Phys.* **42** 125201

[21] Jentschura U D, Mohr P J, Tan J N and Wundt J B 2008 *Phys. Rev. Lett.* **100** 160404

[22] Flannery M R and Oks E 2008 *Eur. Phys. J.* D **47** 27

[23] Devoret M, Girvin S and Schoelkopf R 2007 *Ann. Phys. (Leipzig)* **16** 767

[24] Shytov A V, Katsnelson M I and Levitov L S 2007 *Phys. Rev. Lett.* **99** 246802

[25] Flannery M R and Oks E 2006 *Phys. Rev.* A **73** 013405
[26] Oks E 2004 *Eur. Phys. J.* D **28** 171
[27] Holmlid L 2002 *J. Phys. Condens. Matter* **14** 13469
[28] Oks E 2001 *Phys. Rev.* E **63** 057401
[29] Oks E 2000 *Phys. Rev. Lett.* **85** 2084
[30] Oks E 2000 *J. Phys. B: Atom. Mol. Opt. Phys.* **33** 3319
[31] Carlsen H and Goscinski O 1999 *Phys. Rev. A.* **59** 1063
[32] Lee F, Farrelly D and Uzer T 1997 *Opt. Express* **1** 221
[33] Germann T C, Herschbach D R, Dunn M and Watson D K 1995 *Phys. Rev. Lett.* **74** 658
[34] Cheng C H, Lee C Y and Gallagher T F 1994 *Phys. Rev. Lett.* **73** 3078
[35] Chen L, Cheret M, Roussel F and Spiess G 1993 *J. Phys. B: Atom. Mol. Opt. Phys.* **26** L437
[36] Cortiñas R G, Favier M, Ravon B, Mehaignerie P, Machu Y, Raimond J M, Sayrin C and Brune M 2020 *Phys. Rev. Lett.* **124** 123201
[37] Dutta S K, Feldbaum D, Walz-Flannigan A, Guest J R and Raithel G 2001 *Phys. Rev. Lett.* **86** 3993
[38] Hulet R G, Hilfer E S and Kleppner D 1985 *Phys. Rev. Lett.* **55** 2137
[39] Vainberg V M, Popov V S and Sergeev A V 1990 *Sov. Phys. JETP* **71** 470
[40] Nadezhdin B B and Oks E 1990 *Opt. Spectrosc.* **68** 12
[41] Beletsky V V 2001 *Essays on the Motion of Celestial Bodies* (Basel: Birkhäuser) section 1.7

Chapter 5

Classical analytical description of negative hydrogen ions

Negative hydrogen ions (NHIs), such as, H^- and D^-, have been studied for decades—see, e.g., papers [1–6] and references therein. One example of the practical importance of NHI concerns heating tokamak plasmas by external beams: the employment of NHI beams leads to much better results than the employment of neutral beams. Another example of the practical importance of NHIs relates to high power proton accelerators: the injection of NHI boosts the performance of the accelerators. Besides, NHIs play an essential role in causing opacities of the atmospheres of the A-type stars, as well as in the Sun's atmosphere.

Studies of NHIs are also important from the fundamental point of view. They are the simplest two electron systems. However, due to very significant correlations between the two electrons, NHIs cannot be described by the standard two-electron technique. There is only one bound state in NHIs and it is only weakly bound: the binding energy of the second electron is just about 0.75 eV, which makes this electron analogous to Rydberg electrons.

One of the ways of describing how this bound state is formed is as the bound state of an electron in the field of the electric dipole. In this scenario, the electric dipole moment is provided by the hydrogen or deuterium atom (we recall that in parabolic quantization, the overwhelming majority of states of the hydrogen/deuterium atom has a non-zero average dipole moment), the second electron being attracted to the electric dipole. The system 'electron plus electric dipole' has bound states if the dipole moment is greater than some threshold value—see, e.g., the corresponding *quantum* results in papers [7–11] and references therein.

As for the corresponding *classical* investigations, first we point out the Turner–Fox' paper of 1968 [12], where the authors analyzed the situation for the case of a point dipole. Much later, Howard and Wilkerson [14] in 1995 proceeded from the same setup and utilize the analogy with the system consisting of an electron in the field of two immobile Coulomb centers. The next year, Sastry *et al* [8] reformulated

doi:10.1088/978-0-7503-3939-1ch5

the same physical system as being an electron around a polar molecule. In 2020, Fox's results [13] have been further advanced by Oks [15] who presented a detailed classical analytical study of the oscillatory-precessional motion of the bound electron.

It is important to underline that the results from papers [7–15] are valid only if the outer electron can be considered as a slow subsystem while the hydrogen/deuterium atom can be considered as a rapid subsystem—otherwise the atom cannot be treated as a fixed electric dipole. However, the authors of papers [7–15] did not discussed this validity condition. Another question that remained open was whether the outer electron can be bound by the hydrogen/deuterium atom if the latter is in one of the states of the zero averaged electric dipole moment.

In paper [16], which we follow here, the author dealt with the above questions in frames of classical mechanics. Following paper [16], we analyze a system having the proton (or deuteron) and two Rydberg electrons. One of the electrons (the inner one) is much closer to the nucleus than the other (outer) electron. The inner electron is considered to be in a circular state, so that the electric dipole moment of the pair 'nucleus plus inner electron' is zero.

The inner electron and the nucleus revolve about their center of mass with the frequency

$$\Omega = e/(m_{ier}R_{ie}^{3})^{1/2}. \tag{5.1}$$

In equation (5.1), e is the electron charge, R_{ie} is the radius of the orbit of the inner electron (here and below the subscript 'i.e.' stands for 'inner electron'), and

$$m_{ier} = m_e M_{nucl}/(m_e + M_{nucl}) \tag{5.2}$$

is the reduced mass of the pair 'nucleus plus inner electron'. In equation (5.2), m_e is the electron mass and M_{nucl} is the nuclear mass. For example, if the nucleus is the proton, then $m_{ier} = 0.999456$; if the nucleus is the deuteron, then $m_{ier} = 0.999728$. Here and below we use the atomic units: $\hbar = m_e = e = 1$.

We are interested in the case of the bound motion of the outer electron where its primary frequency ω meets the following condition:

$$\omega << \Omega. \tag{5.3}$$

Later on in this chapter it will be shown that if the binding energy of the outer electron is small enough, condition (5.3) would be fulfilled.

Obviously, if condition (5.3) is met, then the outer electron can be considered as the slow subsystem, and the pair 'nucleus plus inner electron' as the rapid subsystem. The well-known analytical method of separating rapid and slow subsystems prescribes averaging over the rapid subsystem before solving for the slow subsystem.

This results in the following physical situation. The outer electron 'sees' both the nucleus and the inner electron as concentric circular rings having the radii R_{nucl} and R_{ie}, respectively. The nuclear charge is uniformly distributed over the ring of the radius R_{nucl}, while the electron charge is uniformly distributed over the ring of the radius R_{ie}, where

$$R_{ie}/R_{nucl} = M_{nucl}/m_e >>1. \tag{5.4}$$

For example, for the deuteron as the nucleus, one has $R_{ie}/R_{nucl} = 3670$, while for the proton as the nucleus, the ratio is $R_{ie}/R_{nucl} = 1836$.

Since the above two rings do not have the electric dipole moment, then for the outer electron the primary contribution to its effective potential energy is from the quadrupole interaction with the ring corresponding to the inner electron (see, e.g., papers [17–19] and book [20]):

$$U_{ie} = -\,e^2R_{ie}{}^2(3\cos^2\theta - 1)/(4r^3). \tag{5.5}$$

In equations (5.5), r is the absolute value of the radius-vector of the outer electron and θ is the polar angle; the axis of the symmetry of the two rings is chosen as the z-axis.

For the nuclear ring, the corresponding quadrupole contribution is

$$U_{nucl} = e^2R_{nucl}{}^2(3\cos^2\theta - 1)/(4r^3). \tag{5.6}$$

From equations (5.5) and (5.6) it is seen that

$$|U_{ie}/U_{nucl}| = R_{ie}{}^2/(R_{nucl}{}^2) = M_{nucl}{}^2/m_{ie}{}^2 >>1. \tag{5.7}$$

Thus, the nuclear ring contribution can be neglected. Consequently, the total effective potential energy $U(r, \theta)$ becomes

$$U(r, \theta) = -e^2R_{ie}{}^2(3\cos^2\theta - 1)/(4r^3). \tag{5.8}$$

Then the author of paper [16] fixed the angle θ at $\theta = 0$, which is one of the two minima of the effective potential energy from equation (5.8) (the other minimum at $\theta = \pi$). Thereafter, the effective potential energy simplifies to

$$U(z) = -\,A/|z|^3, \qquad A = e^2R_{ie}{}^2/2. \tag{5.9}$$

The expression for the total energy becomes

$$E = (m_{oer}/2)(dz/dt)^2 - A/|z|^3. \tag{5.10}$$

In equation (5.10)

$$m_{oer} = m_e(M_{nucl} + m_e)/(M_{nucl} + 2m_e) \tag{5.11}$$

is the reduced mass of the outer electron. From equation (5.10) one can find that

$$dz/dt = \pm[(2/m_{oer})\,(E + A/|z|^3)]^{1/2}. \tag{5.12}$$

Then the author of paper [16] wrote (here we update the numeration of equations):

'Thus, the effective potential energy from equation (9) corresponds to the attractive potential, in which the outer electron will oscillate back and forth along the axis of the symmetry of the two concentric 'rings' between the following two turning points

$$z_{\pm} = \pm(A/|E|)^{1/3} \tag{5.13}$$

where $|E|$ is the binding energy. The differential equation (5.12) can be integrated analytically yielding the following equation of the motion

$$t = \pm(m_{oer}/2)^{1/2} \int dz/(A/|z|^3 - |E|)^{1/2} = \pm(m_{oer}/2)^{1/2}(A^{1/3}/|E|^{5/6}) \int u^{3/2}du/(1 - u^3)^{1/2} \quad (5.14)$$

where we denoted

$$u = z/z_+. \quad (5.15)$$

The integral in the utmost right side of equation (5.14) can be expressed through elliptical functions. However, the resulting expression is too bulky, so that we do not reproduce it here.'

It was possible obtaining the following analytical result for the period T of the oscillations of the outer electron

$$\mathrm{T} = 2^{3/2}\mathrm{m}_{oer}{}^{1/2}(\mathrm{A}^{1/3}/|\mathrm{E}|^{5/6})\int_0^1 \mathrm{u}^{3/2}\mathrm{du}/(1-\mathrm{u}^3)^{1/2} = [2(2\pi\Gamma(1/3^{1/2}\Gamma(5/6)/\Gamma(1/3)](\mathrm{A}^{1/3}/|\mathrm{E}|^{5/6}), \quad (5.16)$$

where $\Gamma(w)$ is the gamma-function. For the frequency ω of the oscillations, the corresponding expression is

$$\omega = [(\pi/2)^{1/2}\Gamma(1/3)/\Gamma(5/6)]\, |E|^{5/6}/(m_{oer}{}^{1/2}A^{1/3}) = \{\pi^{1/2}\Gamma(1/3)/[2^{1/6}\Gamma(5/6)]\} |E|^{5/6}/(m_{oer}{}^{1/2}e^{2/3}R_{ie}{}^{2/3}). \quad (5.17)$$

For deriving equation (5.17), the author of paper [16] utilized the definition of the quantity A from equation (5.9). The radiation by the outer electron occurs at the primary frequency ω given by equation (5.17).

By combining equations (5.1) and (5.17), the authors of paper [16] obtained the ratio of the frequencies:

$$\omega/\Omega = (\pi^{1/2}/2^{1/6})[\Gamma(1/3)/\Gamma(5/6)](m_{ier}/m_{oer})^{1/2}[|E|R_{ie}/e^2]^{5/6}. \quad (5.18)$$

From equation (5.18) one can see that the binding energy of the outer electron should be small enough for the validity condition, given by equation (5.3) to be met:

$$|E| << |E|_{crit} = (2/\pi^3)^{1/5}[\Gamma(5/6)/\Gamma(1/3)]^{6/5}\, e^2/R_{ie} \approx 0.41\, |E|_{ie} \quad (5.19)$$

($m_{oer} = m_{ier} = m_e$ has been assumed for simplicity). Here the quantity $|E|_{ie} = e^2/(2R_{ie})$ is the inner electron binding energy.

The authors of paper [16] also analytically did a relativistic study of this system, assuming $m_{oer} = m_{ier} = m_e$ for simplicity. The relativistic expression for the total energy (accounting also for the rest mass of the electron) was

$$E_t = m_e c^2/[1 - (dz/dt)^2/c^2]^{1/2} - A/|z|^3. \quad (5.20)$$

From equation (5.20) he obtained the relativistic formula for the turning points (by setting $dz/dt = 0$):

$$z_{\pm} = \pm[A/(m_e c^2 - E_t)]^{1/3} = \pm[A/|E|]^{1/3}. \tag{5.21}$$

In equation (5.21)

$$|E| = m_e c^2 - E_t \tag{5.22}$$

is the outer electron binding energy.

From equation (20) he got

$$dz/dt = \pm c\left\{1 + [m_e c^2/(m_e c^2 + A/|z|^3 - |E|)]^2\right\}^{1/2}. \tag{5.23}$$

Then the author of paper [16] scaled the binding energy to the following dimensionless form:

$$\varepsilon = |E|/(m_e c^2). \tag{5.24}$$

By employing the notations from equations (5.15) and (5.24), [16], he managed to analytically integrate the differential equation (5.23) and obtained the following equation of the motion:

$$t = \pm (z_+/c) \int \mathrm{d}u/\left\{1 + [1/(1 - \varepsilon + \varepsilon/u^3)]^2\right\}^{1/2}. \tag{5.25}$$

He noted: 'The integral in the right side of equation (25) can be expressed through the Appell hypergeometric function of two variables.'

The expression for the period of the outer electron oscillations is

$$T = (4z_+/c) \int_0^1 \mathrm{d}u/\left\{1 + [1/(1 - \varepsilon + \varepsilon/u^3)]^2\right\}^{1/2}. \tag{5.26}$$

After the integration, equation (5.26) yields

$$T = (4z_+/c)\, I(\varepsilon) \tag{5.27}$$

where

$$\begin{aligned} I(\varepsilon) = {} & F_1[1/3,\ 1/2,\ 1/2,\ 4/3,\ (2 - 2\varepsilon + \varepsilon^2)/(\varepsilon^2 - \varepsilon - i\varepsilon), \\ & (2 - 2\varepsilon + \varepsilon^2)/(\varepsilon^2 - \varepsilon + i\varepsilon)] + \\ & [(1/\varepsilon - 1)/4]\, F_1[4/3,\ 1/2,\ 1/2,\ 7/3,\ (2 - 2\varepsilon + \varepsilon^2)/(\varepsilon^2 - \varepsilon - i\varepsilon), \\ & (2 - 2\varepsilon + \varepsilon^2)/(\varepsilon^2 - \varepsilon + i\varepsilon)]. \end{aligned} \tag{5.28}$$

In equation (5.28), F_1 are the Appell hypergeometric functions of two variables. It should be emphasized that equation (5.28) yields real values despite the appearance of the complex unity '*i*'.

Figure 5.1 presents the dependence of the quantity I on the scaled energy ε. One can see that the dependence is relatively weak. The analytical results show that the quantity I grows only from $1/2^{1/2}$ to 1, as the scaled energy ε changes from 0 to ∞.

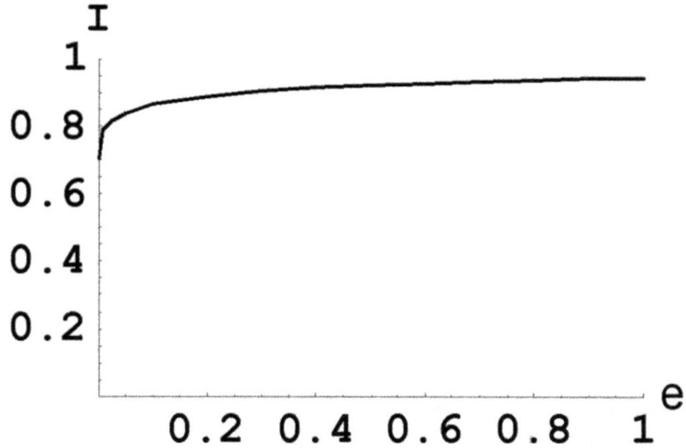

Figure 5.1. Plot of the function $I(\varepsilon)$ entering the expression for the period of the oscillations of the outer electron (equations (5.27) and (5.28)). Reprinted from [16] with permission from the Canadian Journal of Physics.

The corresponding frequency of the oscillations, i.e., the primary frequency at which the outer electron emits the electromagnetic radiation, is

$$\omega = \left\{\pi c/[2^{2/3}I(\varepsilon)]\right\}[|E|/(e^2R_{ie}^2)]^{1/3}. \quad (5.29)$$

This result can be utilized for the experimental detection of this physical system.

By combining equationss (5.1) and (5.29), the author of paper [16] obtained the following ratio of the frequencies

$$\omega/\Omega = \left\{\pi/[2^{2/3}I(\varepsilon)]\right\}(m_{ie}rc^2)^{1/2}|E|^{2/3}R_{ie}^{5/3}/e^{10/3}. \quad (5.30)$$

It is seen that if the outer electron binding energy $|E|$ is small enough, the validity condition (5.3) is met, that is, if

$$|E| << |E|_{crit} \sim e^5/[R_{ie}^{5/2}(m_ec^2)^{3/2}] \sim |E|_{ie}^{5/2}/(m_ec^2)^{3/2} \quad (5.31)$$

where again $|E|_{i.e.}$ is the inner electron binding energy.

The author of paper [16] underscored the following:

'The states that we found for the above systems could be considered as classical counterparts of the double-excited states of NHI, the quantum studies of which are presented, e.g. in papers [5, 6]. The existence of classical counterparts of the double-excited states of NHI is a *counterintuitive* result.'

He further emphasized the following important fact.

'In the quantum method employing the $1/n$-expansion (n being the principal quantum number), the primary term corresponds to the classical description of the system—see, e.g. paper [21] and references therein. The relative error (i.e., the quantum correction) is $\sim 1/(\pi n)^2$—see, e.g., the textbook [22, section 3.1]. In the system under consideration we have $n^2 \sim R_{ie}/a_0$, where a_0 is the Bohr radius. Therefore, already for $n \sim 3$ (i.e., for $R_{ie}/a_0 \sim 10$), the relative quantum correction is $\sim 10^{-2}$ and therefore can be neglected (and even more so for $n \sim 10$ or higher).

References

[1] Baskerville A L, King A W and Cox H 2019 *R. Soc. Open Sci.* **6** 181357
[2] Faircloth D and Lawrie S 2018 *New J. Phys.* **20** 025007
[3] Bacall M and Wada M 2015 *Appl. Phys. Rev.* **2** 021305
[4] Balling P, Andersen H H, Brodie C A, Pedersen U V, Petrunin V V, Raarup M K, Steiner P and Andersen T 2000 *Phys. Rev.* A **61** 022702
[5] Bürgers A and Lindroth E 2000 *Eur. Phys. J.* D **10** 327
[6] Rau A R P 1996 *J. Astrophys. Astron.* **17** 113
[7] Connolly K and Griffiths D J 2007 *Am. J. Phys.* **75** 524
[8] Sastry G P, Srinivas V and Madhav A V 1996 *Eur. J. Phys.* **17** 275
[9] Turner J E 1977 *Am. J. Phys.* **45** 758
[10] Fox K and Turner J E 1966 *J. Chem. Phys.* **45** 1142
[11] Fox K and Turner J E 1966 *Am. J. Phys.* **34** 606
[12] Turner J E and Fox K 1968 *J. Phys. A (Proc. Phys. Soc.)* **1** 118
[13] Fox K 1968 *J. Phys. A (Proc. Phys. Soc.)* **1** 124
[14] Howard J E and Wilkerson T D 1995 *Phys. Rev.* A **52** 4471
[15] Oks E 2020 *Symmetry* **12** 1275
[16] Oks E 2021 *Can. J. Phys.* **99** 9
[17] Oks E 2020 *Can. J. Phys.* **98** 857
[18] Oks E 2021 *New Astron.* **84** 101500
[19] Oks E 2020 *New Astron.* **74** 101301
[20] Oks E 2020 *Analytical Advances in Quantum and Celestial Mechanics: Separating Rapid and Slow Subsystems* (Bristol: IOP Publishing) chs 3 and 4
[21] Vainberg V M, Popov V S and Sergeev A V 1990 *Sov. Phys. JETP* **71** 470
[22] Migdal A B 2000 *Qualitative Methods in Quantum Mechanics* (Boca Raton, FL: CRC Press)

IOP Publishing

Advances in Physics of Rydberg Atoms and Molecules

Eugene Oks

Chapter 6

Radio lines emitted by Rydberg hydrogen atoms in the Universe: classical calculations of new data for their Stark broadening

Radio recombination lines are widely used for analyzing a variety of processes in the H II regions (planetary nebulas)—see, e.g., [1, 2]. The leading role in the Stark broadening of radio recombination lines is played by inelastic collisions with electrons and protons [3, 4].

In paper [5], which we briefly overview in this chapter, the authors were motivated by the following situation. There was a very significant discrepancy—reaching hundreds of percent—in the calculations of cross-sections of inelastic collisions of Rydberg atoms with electrons between the results of papers [3, 6–8]. It translates into a very large discrepancy in calculations of the widths of radio recombination lines.

It should be emphasized that the calculations from [3, 6], which most astrophysicists use for deriving the electron density N_e from the observed widths, employed the *perturbation theory and the dipole approximation*. Therefore, those calculations are not valid for relatively small velocities/energies of the perturbing electrons. For this reason, one of the goals set by the authors of paper [5] was to determine whose results for the cross-sections of inelastic collisions were the most accurate.

Besides, the calculations from [3, 6–8] disregarded such processes as the ionization and (for collisions with protons) the charge exchange. Therefore, another goal set by the authors of paper [5] was to take into account all major processes, pertinent to the Stark broadening of radio recombination lines, within a single theoretical framework.

For achieving these goals, in paper [5] there was utilized the classical path Monte Carlo method [9–12]. The method used by the authors of paper [5] was non-perturbative and did not employ the dipole approximation. This allowed them to determine which of the corresponding results by other authors were the most reliable. Some details are presented below.

doi:10.1088/978-0-7503-3939-1ch6

The authors of paper [5] performed a numerical solution of the Hamilton equations for the ABC-system, where C was the atomic electron, B the hydrogen nucleus, and A the perturbing charged particle (electron or proton). The impact parameter for the collision of the electron or proton with the hydrogen atoms, the momentum of the atomic electron and its spatial orientation were randomly chosen by the authors of paper [5] consistent with the Monte Carlo method.

The authors of paper [5] wrote, in particular:

'A collision between A and the hydrogen atom BC can lead to one of four processes, of which the *first three are inelastic*:

(a) ionization;
(b) charge exchange (if particle A is proton);
(c) collision changing the principal quantum number n_i of the atomic electron;
(d) elastic collision (n_i remains the same).

For chosen n_i and an initial energy E_A of particle A, our code calculated $(3 - 5) \times 10^4$ trajectories by numerically integrating equation (6.1). We defined the collision as inelastic (process (c)) if $|n_f - n_i| \geqslant 0.5$, where n_f is the final principal quantum number. Otherwise, the collision was considered as elastic. The principal quantum number was defined as $n = a^{1/2}$, where a is the major semi-axis of the elliptical orbit of the atomic electron (expressed in atomic units). The cross-section of collisions leading to either process (a), or (b), or (c), was calculated by the formula:

$$\sigma = (N_R/N)\pi b_{\max}^2 . \qquad (.6.1)$$

Here N is the total number of trajectories calculated for given n_i and E_A, N_R is the total number of trajectories resulting in either process (a), or (b), or (c).'

Figure 6.1 presents the results of the calculations for the inelastic collision of the perturbing electron with a hydrogen atom in the state of $n_i = 30$, the outcome of the collision being the transition to the state of $n_f = 25$ (process (c)).

Figure 6.2 presents the results of the calculations for the inelastic collision of the perturbing electron with a hydrogen atom in the state of $n_i = 30$, the outcome of the collision being the ionization (process (a)).

Figure 6.3 demonstrates how the probability $W_q(b)$ of the transition from the state of $n_i = 100$ to one of adjacent states n_f ($|n_f - n_i| \leqslant 3$) depends on the scaled impact parameter b/n_i^2 for electronic collisions. The quantity $W_q(b)$ was defined as follows:

$$W_q(b) = 0.5 \times [W(100 \to 100 + q) + W(100 \to 100 - q)]. \qquad (6.2)$$

One can see first, that the maximum of the probability W_q moves to smaller b, as E_e grows. Second, at the fixed value of E_e, the maximum of the probability W_q moves to smaller b, as q grows.

The obtained dependence $W_q(b)$ enabled the authors of paper [5] to calculate the cross-section of the inelastic transition $n \to n + q$: $\sigma_{n\to n+q} = 2\pi\int_0^\infty W_q(b)bdb$.

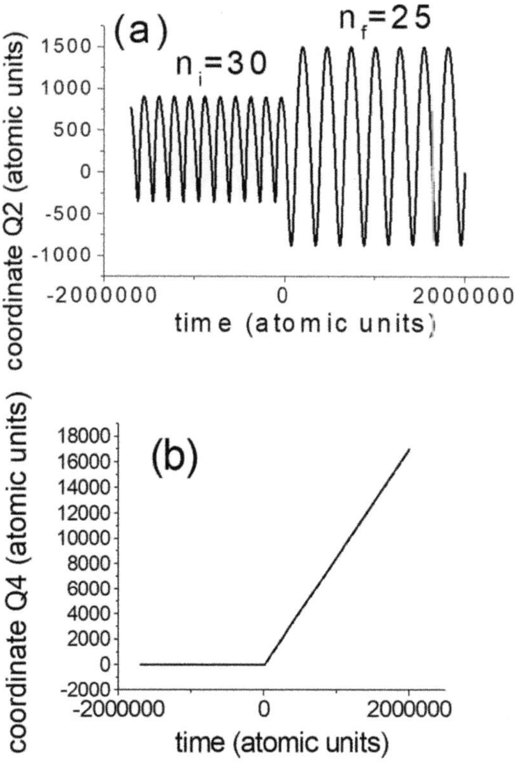

Figure 6.1. Evolution of one of the coordinates of the atomic electron (a) and of one of the coordinates of the incident electron (b) during an inelastic collision of the latter with a hydrogen atom, resulting into the transition of the atomic electron to the state of lower principal quantum number. The closest approach corresponds to the instant $\tau \approx 0$. Reprinted from [5].

Table 6.1 demonstrates the comparison of the results from paper [5] to the corresponding results by other authors for the cross-section of inelastic collisions of a hydrogen atom in the state of $n_i = 100$ with a perturbing electron. The results from paper [5] are presented in line α. Line β contains the results obtained by using equations from book [7], the results being derived semiclassically. Line γ presents the results corresponding to equation (478) from book [6]. Concerning the latter, the authors of paper [5] noted that the applicability of equation (478) from book [6] was 'limited by the employment of both the perturbation theory and the dipole approximation.' Further, the authors of paper [5] wrote: 'Besides, in his formula (478), instead of the square of the orbital mometum l^2 of the atomic electron, an average value $<l^2>_{av} = n^2/2$ was used. We also note that at relatively small energies E_e, in his formula (478) the minimal impact parameter $\rho_{min} \approx n^2\hbar/m_e V_e$ exceeds the maximal impact parameter $\rho_{max} \approx V_e/\omega_{n,n-1}$ (V_e is the velocity of the incident electron) and the formula becomes inapplicable.'

Line δ displays the corresponding semi-empirical cross-sections from table 6.1 of paper [8]. Line ε presents the corresponding results calculated by equations from paper [7] obtained in the Born approximation.

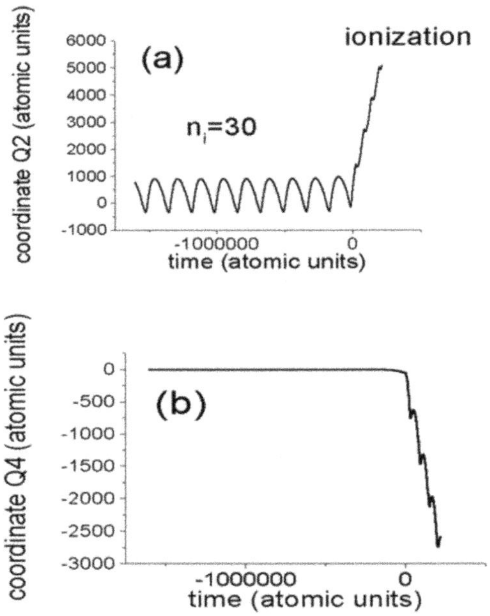

Figure 6.2. Same as in figure 6.1, but for an inelastic collision leading to the ionization of the hydrogen atom. The closest approach corresponds to the instant $\tau \approx 0$. Reprinted from [5].

Further, the authors of paper [5] specified the following.

'The cross-sections presented in lines α, β, γ, and ε allow for transitions to all levels n_f, where $n_f \neq n_i$. The cross-sections presented in line δ allow only for transitions to the six nearest levels $n_f = n_i \pm 1$, $n_i \pm 2$, $n_i \pm 3$; however, the contribution of the transitions to further levels n_f ($|n_f - n_i| > 3$) is relatively small.

Our calculations are non-perturbative and do not use the dipole approximation. By benchmarking the results from [6–8] to our calculations, we found that the most accurate are the semi-empirical formulas for the cross-sections by Gee *et al* [8].

Further, for calculating the Stark width (FWHM) of hydrogen radio recombination lines, one can use the above dependence of the inelastic cross-section σ on the energy/velocity of the incident charged particles and obtain the FWHM by the formula: $\Gamma = 2N < V\sigma >$. Here N is the density of the charged particles, V—their velocity relative to the hydrogen atom.'

Figure 6.4 presents the results from paper [5] the dependence of the cross-section of inelastic collisions of protons or electrons with a hydrogen atom on the initial principal quantum number n_i. The energy of the particle was assumed to be 1 eV.

Figure 6.4 illuminates that for the inelastic collisions of a hydrogen atom in the state of n_i greater than or equal to 500 with protons or electrons, the cross-section of protonic collisions can be much larger than the cross-section of the electronic collisions, provided that in both cases the energy of the perturbing particle is the same. Here is a physical explanation by the authors of paper [5] (where we update the numeration of figures and equations).

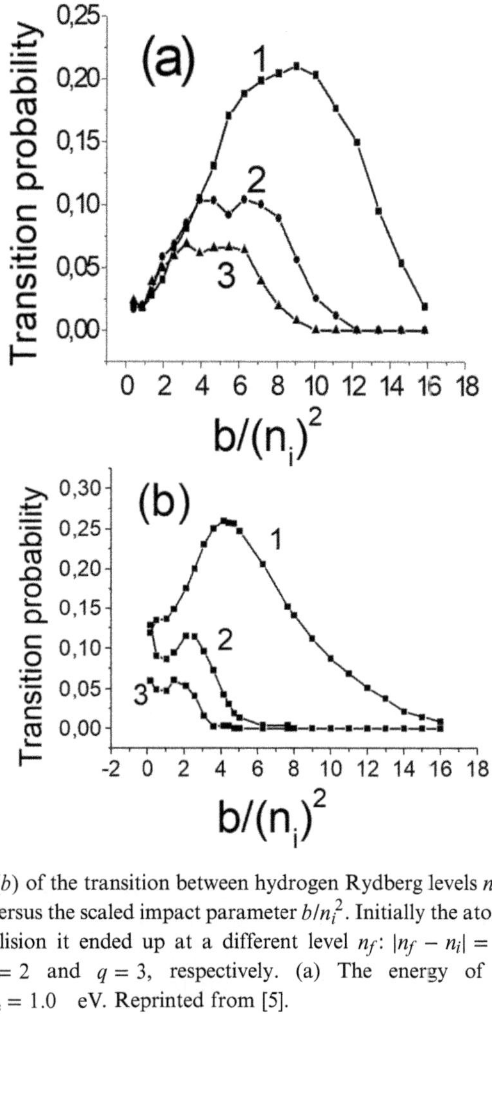

Figure 6.3. Probability $W_q(b)$ of the transition between hydrogen Rydberg levels n_i и n_f, caused by a collision with an incident electron, versus the scaled impact parameter b/n_i^2. Initially the atomic electron was at the level of $n_i = 100$. After the collision it ended up at a different level n_f: $|n_f - n_i| = q$. The curves 1, 2, and 3 correspond to $q = 1$, $q = 2$ and $q = 3$, respectively. (a) The energy of the incident electron is: (a) $E_e = 0.056$ eV; (b) $E_e = 1.0$ eV. Reprinted from [5].

Table 6.1. Cross-sections for transitions $n_i \to n_f$ in hydrogen atoms at various energies of the incident electrons; $n_i = 100$. Reprinted from [5].

E_e(eV)		0.054	5.44	136.06
$10^{-8}\sigma$ (atomic units)	α	280.2	47.8	2.7
	β	121.6	44.9	3.3
	γ	N/A ($\sigma < 0$)	34.7	12.3
	δ	273.5	52.4	3.4
	ε	3518.0	65.0	3.5

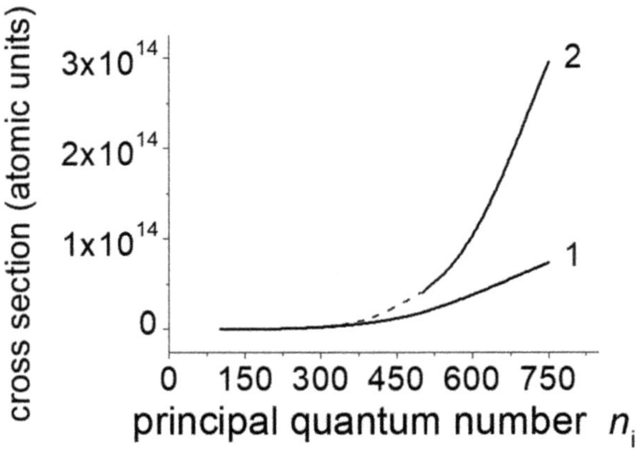

Figure 6.4. Cross-section of inelastic collisions of a hydrogen atom with either electrons (curve 1) or protons (curve 2) charged of the energy $E = 1$ eV versus the initial principal quantum number n_i. Reprinted from [5].

'An inelastic collision with a charged particle is effective if

$$V/\rho_{\text{eff}} > \omega_{n+1,n} = 1/n^3. \tag{6.3}$$

Here V/ρ_{eff} is the characteristic frequency of variation of the electric field of the incident particle (V is its velocity, ρ_{eff} is the effective impact parameter), $\omega_{n+1,n}$ is the separation between the levels of the principal quantum numbers $n+1$ and n. From figure 6.3 it can be inferred that the effective impact parameter is $\rho_{\text{eff}} = kn^2$, where the coefficient $k \sim 5$–10. Therefore, equation (6.3) can be re-written as

$$n > k/V. \tag{6.4}$$

Consequently, in the case of equal energies of the incident electron and of the incident proton, protonic collisions become effective at significantly larger n than electronic collisions. Combined with the fact that the inelastic cross-section increases when the velocity of the charged particle decreases (this can be seen from table 6.1), this explains the result presented in figure 6.4.

In summary, the method used by the authors of paper [5] makes it possible to calculate the corresponding cross-sections of inelastic collisions for a broad range of velocities of perturbing electrons or protons. This includes the most difficult range of energies where the velocity of the atomic electron is of the same order of magnitude as the velocity of the perturbing particle. Therefore, the results from paper [5] have a practical significance for studying atomic processes in astrophysical and/or laboratory plasmas.

References

[1] Gordon M A and Sorochenko R L 2002 *Radio Recombination Lines: Their Physics and Astronomical Applications* (Dordrecht: Kluwer)

[2] Mosse C, Calisti A, Stamm R, Talin B, Bureyeva L A and Lisitsa V S 2004 *J. Phys. B: At. Mol. Opt. Phys.* **37** 1343

[3] Griem H R 1967 *Astrophys. J.* **148** 547
[4] Minaeva L A and Sobelman I I 1968 *J. Quant. Spectr. Rad. Transfer* **8** 783
[5] Gavrilenko V P and Oks E 2007 *Phys. Scripta* **76** 43
[6] Griem H R 1974 *Spectral Line Broadening by Plasmas* (New York, London: Academic)
[7] Lebedev V S and Beigman I L 1998 *Physics of Highly Excited Atoms and Ions* (Berlin: Springer)
[8] Gee C S, Percival I C, Lodge J C and Richards D 1976 *Mon. Not. R. Astr. Soc.* **175** 209
[9] Karplus M, Porter R N and Sharma R D 1965 *J. Chem. Phys.* **43** 3259
[10] Abrines R and Percival I C 1966 *Proc. Phys. Soc.* **88** 861
[11] Olson R E and Salop A 1977 *Phys. Rev.* A **16** 531
[12] Vrinceanu D 2005 *Phys. Rev.* A **72** 022722

IOP Publishing

Advances in Physics of Rydberg Atoms and Molecules

Eugene Oks

Chapter 7

Oscillatory-precessional motion of a Rydberg electron around a polar molecule

In atomic and molecular physics, a hydrogen atom is the first most fundamental problem. The second most fundamental problem is an electron in the field of an electric dipole. This is especially true for the situation where the distance of the electron from the dipole is much larger than the dimension of the dipole, i.e., in the situation where the dipole can be considered point-like. There were lots of theoretical studies of this fundamental problem (see, e.g., paper [1] and references therein)—starting from Fermi–Teller's paper of 1947 [2].

For Rydberg states of this system, the classical description is appropriate. Turner and Fox [3] provided an analytical study of this problem in the elliptical coordinates (ξ, η, φ). They demonstrated that the classical bound motion is possible in some region ($\xi_{min} < \xi < \xi_{max}$, $\eta_{min} < \eta < \eta_{max}$) for any finite size of the dipole. The motion is confined within a torus formed by the revolution of the two limiting ellipses, corresponding to $\xi = \xi_{min}$ and $\xi = \xi_{max}$, and of the two limiting parabolas, corresponding to $\eta = \eta_{min}$ and $\eta = \eta_{max}$—the revolution about the dipole axis.

Kryukov and Oks [4] started by analyzing circular orbits of a negative charge in the field of a finite dipole in the cylindrical coordinates—in the situation where the plane of the orbit of the negative charge is perpendicular to the dipole axis. (It should be noted that in paper [4] the finite dipole was considered to be the configuration of stationary proton and electron, but the analytical results obtained in paper [4] are relevant to the motion of an electron in the field of any finite dipole.) The authors of paper [4] demonstrated the existence of stable circular orbits for the case where the dipole size exceeds (or is equal to) some critical value—the value that was found to be equal to the corresponding quantum critical value from many quantal studies of this system—see, e.g., paper [2] and references therein. Moreover, Kryukov and Oks [4] went beyond the case of circular orbits of the electron: they provided an analytical description of a stable *conic-helical* motion of the electron in the field of

doi:10.1088/978-0-7503-3939-1ch7

the finite dipole: specifically, the orbit was a helix confined to the surface of a right frustum of a cone, whose axis coincided with the dipole axis.

From the physical point of view, the possibility of deriving these analytical results in papers [3, 4] was due to the fact that this physical system is a particular case of a more general system: a charged particle in the field of two stationary Coulomb centers. Both this general system and its particular case analyzed in papers [3, 4] possess higher than geometrical symmetry. The manifestation of this algebraic symmetry is the existence of an additional conserved quantity. This conserved quantity is the projection of the super-generalized Runge–Lenz vector on the axis connecting the stationary charges [5].

Now let us enumerate theoretical studies devoted to the classical description of the motion of a charged particle in the field of a *point-like* electric dipole. Fox [6] provided the first analytical description of this problem. He demonstrated that for the bound motion to be possible, it should be confined to a sphere and correspond to the energy $E = 0$. Fox [6] presented his results in the form of integrals, but he did not perform the integration and he did not provide a qualitative description of the motion (more details from paper [6] are given in the next part of this chapter).

Jones [7] obtained analytical results for the orbits that are confined to a semicircle along the meridian of the sphere. In his study, the particle was assumed to be positively charged. For this case he noted that the particle motion is mathematically identical to the pendulum motion—because the potential of the dipole at the constant radius is mathematically equivalent to the potential of a simple pendulum.

McDonald [8] considered two kinds of orbits. One kind was the same semicircular orbit along the meridian studied by Jones [7]. Another kind was an orbit representing a circle corresponding to some specific parallel of latitude. He reproduced some of Fox's results [6] (though without any reference to Fox's paper [6]). He also made a note that, generally speaking, the motion would be represented by large oscillations with respect to the polar angle θ complemented by a slow precession about the axis of the dipole.

In the present chapter we follow paper [9]. In paper [9] the author started where Fox stopped in his paper [6]. In that paper it was shown that generally the oscillations of the electron along a meridian on a sphere (θ-oscillations) and the precession along a parallel of latitude (the precession along the coordinate φ) can happen on the same time scale (thus, the corresponding statement by McDonald [8] was incorrect).

In paper [9], for the general case there was derived the dependence $\varphi(\theta)$. It was derived in the form of a one-fold integral. Then for the case where the projection p_φ of the angular momentum on the axis of the electric dipole is relatively small, there was obtained an explicit analytical result for the dependence of the angular variable φ on the other angular variable θ.

Next, for the case of $p_\varphi = 0$, there was derived an explicit analytical result for the dependence of the time t on the angular variable θ. Then, for the situation where the projection p_φ of the angular momentum is equal to its maximum possible (for the bound motion) value, there was obtained an explicit analytical expression for the period of the revolution of the electron along the angle φ, (i.e., along the parallel of latitude).

Next, the author of paper [9] derived analytically (as one-fold integrals) the time evolution of the angle θ, the period T_θ of the meridional oscillations (i.e., oscillations with respect to the angle θ, and the change of the angle φ during one half-period of the meridional oscillations. Then for the case where the projection p_φ of the angular momentum on the axis of the electric dipole is relatively small, there were obtained the corresponding explicit analytical results.

Lastly, in paper [9] there were studied conditions, under which the trajectory of the electron would become a closed curve (i.e., the conditionally-periodic motion would become periodic). Below are all the details.

As in Fox's paper [6], the author of paper [9] dealt with an electric dipole of the dipole moment D. The dipole is located at the origin. The motion of an electron was studied in spherical polar coordinates (r, θ, φ). The z-axis of this reference frame was chosen along the dipole axis—specifically, in such a way that the positive pole points to the upper hemisphere. Then the energy can be represented in the form:

$$E = m[(dr/dt)^2 + r^2(d\theta/dt)^2 + r^2 \sin^2\theta\,(d\varphi/dt)^2]/2 - eD\cos\theta/r^2. \quad (7.1)$$

In equation (7.1), m and e are the electron mass and the absolute value of the electron charge.

The z-projection of the electron angular momentum M_z is conserved, as the consequence of the axial symmetry. This can be also seen from the fact that the energy does not depend on φ.

In paper [6] Fox used the notation p_φ for the z-projection of the electron angular momentum. This was because p_φ is actually the generalized momentum corresponding to the dynamical variable φ:

$$p_\varphi = mr^2 \sin^2\theta\,(d\varphi/dt) = \text{const}. \quad (7.2)$$

Thus, the symmetry of the problem allows separating the θ- and φ-motions. We recall that r = const, as a requirement for the bound motion to exist, so that only θ and φ are the dynamic variables.

Fox [6] obtained the following equation for the

$$(dx/dt)^2 = [2eD/(mr^4)](-x^3 + x - K). \quad (7.3)$$

In equation (7.3),

$$x = \cos\theta, \quad K = p_\varphi^{\,2}/(2meD). \quad (7.4)$$

From equation (7.2) it is easy to obtain the corresponding differential equation for the φ-motion:

$$d\varphi/dt = p_\varphi/(mr^2 \sin^2\theta) = p_\varphi/[mr^2(1 - x^2)]. \quad (7.5)$$

After obtaining $x(t)$ by solving equation (7.3), it would be possible to get $\varphi(t)$ from equation (7.5).

For the existence of the bound motion, the polynomial

$$y(x) = -x^3 + x - K \quad (7.6)$$

in equation (7.3) must be positive in some range of x within the interval from—1 to 1 (as noted by Fox [6]). This polynomial has a maximum equal to $-K + 2/3^{3/2}$ at

$x = 1/3^{1/2}$ (as well as a negative minimum equal to $-K - 2/3^{3/2}$ at $x = -1/3^{1/2}$). The maximum should be positive (otherwise the bound motion would not exist), so that it is required (Fox [6])

$$K < 2/3^{3/2} = K_{\max}. \tag{7.7}$$

In view of equation (7.4), the condition (7.7) can be rewritten as

$$D > 3^{3/2}p_\varphi^{\ 2}/(4me). \tag{7.8}$$

If the condition (7.7) is satisfied, then the x-motion would be confined between two positive turning points. This means that the bound motion would take place in the upper hemisphere (Fox [6]).

Now we present here the results obtained in paper [9]. Figure 7.1 displays the dependence of the polynomial $y(x, K)$ on its arguments in the range of K from 0 to $K_{\max} = 2/3^{3/2}$ and in the range of x varies from –1 to 1. Figure 7.1 shows that in this range of K, the maximum value of y is positive. Consequently, there is actually a range of x for the classical bound motion.

In paper [9], positive turning points were denoted as x_2 and x_3 ($x_3 > x_2$). The values x_2 and x_3 are the real roots of the cubic equation:

$$x^3 - x + K = 0. \tag{7.9}$$

The author of paper [9] solved equation (7.9) and found the explicit formulas for the turning points:

$$\begin{aligned} x_2(K) = {} & (-1)^{2/3}2^{1/3} / [(729K^2 - 108)^{1/2} - 27K]^{1/3} - \\ & (-1)^{1/3}[(729K^2 - 108)^{1/2} - 27K]^{1/3} / (2^{1/3}3) \end{aligned} \tag{7.10}$$

$$\begin{aligned} x_3(K) = {} & 2^{1/3} / [(729K^2 - 108)^{1/2} - 27K]^{1/3} \\ & - [(729K^2 - 108)^{1/2} - 27K]^{1/3} / (2^{1/3}3). \end{aligned} \tag{7.11}$$

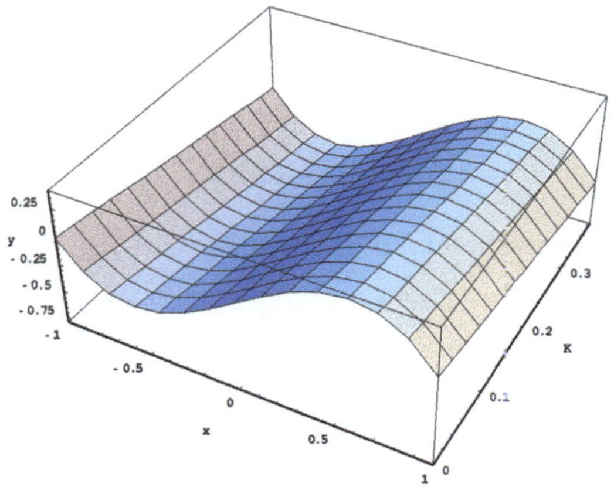

Figure 7.1. Three-dimensional plot of the polynomial $y(x, K)$ from equation (4.6). Reprinted from [9] with permission from MDPI.

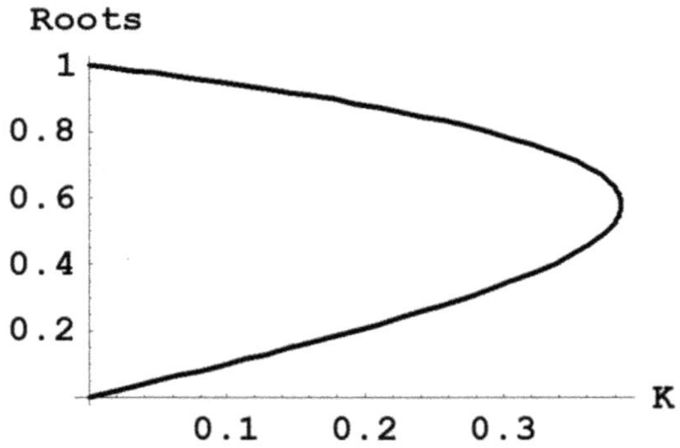

Figure 7.2. Plot of both positive roots x_2 and x_3 of the cubic equation (7.8) versus $K = p_\varphi{}^2 / (2meD)$:.$K = K_{max} = 2/3^{3/2}$. At $K = K_{max}$, one has $x_2(2/3^{3/2})$ and $x_3(2/3^{3/2}) = 1/3^{1/2}$. $x_2(K)$ and $x_3(K)$ are the lower part and the upper part of the double-valued curve, respectively. Reprinted from [9] with permission from MDPI.

Figure 7.2 presents a plot of the roots $x_2(K)$ and $x_3(K)$. In figure 7.2, $x_3(K)$ and $x_2(K)$ are the upper and the lower parts of the double-valued curve, respectively. The upper and lower parts intersect at

In paper [9] there was introduced a scaled dimensionless time τ:

$$\tau = t[2eD/(mr^4)]^{1/2}. \tag{7.12}$$

Then equation (7.3) can be rewritten as follows:

$$d\tau = \pm\, dx / (-x^3 + x - K)^{1/2}. \tag{7.13}$$

The author of paper [9] noted that Fox [6] chose only the minus sign in his equation (7.12) analogous to equation (7.13) and that actually both signs should be taken into account, as explained below.

Next in paper [9] there was a study of limiting cases. In the situation where o $K = K_{max} = 2 / 3^{3/2}$, the electron moves in a circle along the parallel of latitude corresponding to $\cos\theta = 1/3^{1/2}$ (no θ-motion). The value of $\cos\theta = 1/3^{1/2}$ corresponds to $\theta = 0.9553$ rad $= 54.74$ degrees. According to equation (7.5), the electron moves around this circle with the constant angular velocity

$$d\varphi/dt = 3p_\varphi/(2mr^2). \tag{7.14}$$

This angular velocity corresponds to the period

$$T = 4\pi mr^2/(3p_\varphi). \tag{7.15}$$

According to equation (7.4): $p_\varphi = (2KmeD)^{1/2}$. Consequently, for $K = K_{max} = 2/3^{3/2}$ one has $p_\varphi = 2(meD)^{1/2}/3^{3/4}$. Therefore, equations (7.14) and (7.15) can be reformulated in the following way:

$$d\varphi/dt = 3^{1/4}[eD/(mr^4)]^{1/2}, \quad T = (2\pi/3^{1/4})[mr^4/(eD)]^{1/2}. \tag{7.16}$$

In the opposite situation where $K \ll 1$, equations (7.9) and (7.10) reduce to the following:

$$x_2(K) \approx K, \quad x_3(K) \approx 1 - K/2. \tag{7.17}$$

In the subcase where $K = 0$ (meaning that $p_\varphi = 0$), the electron oscillates along a semicircle located in a meridional plane in the upper hemisphere (no φ-motion). In paper [9] this subcase was studied in more detail before returning to the general situation.

Namely, after integrating equation (7.13) analytically, the following explicit dependence of the scaled time τ on x (i.e., the dependence of τ on $\cos\theta$) has been obtained

$$\tau = -\{\pm 2i\, \mathbf{F}[\arcsin(-x)^{1/2}, -1]\}. \tag{7.18}$$

In equation (7.18), $\mathbf{F}(\alpha, q)$ is the incomplete elliptic integral of the first kind. It should be emphasized that the right side of equation (7.18) is in fact real for the range of x from 0 to 1 (i.e., for the range where the motion occurs)—despite the formal appearance of the imaginary unit i. For $x \ll 1$, equation (7.18) simplifies as follows:

$$\tau \approx \pm 2x^{1/2}. \tag{7.19}$$

Figure 7.3 shows the time evolution of the electron as it moves from a point on the equator ($x = 0$ at $\tau = 0$) to the north pole ($x = 1$ at $\tau = 2.622$), then changes its geographical longitude in the upper hemisphere by 180 degrees and moves to the equator ($x = 0$ at $\tau = 5.244$), then back to the north pole ($x = 1$ at $\tau = 7.866$), and then again change its geographical longitude in the upper hemisphere by 180 degrees and moves to the equator ($x = 0$ at $\tau = 10.488$). We note that in paper [9], the corresponding time evolution was not described accurately.

The full period of oscillations is equal to

$$\tau_0 = 4\tau(1) = 10.488 \tag{7.20}$$

or in the usual units

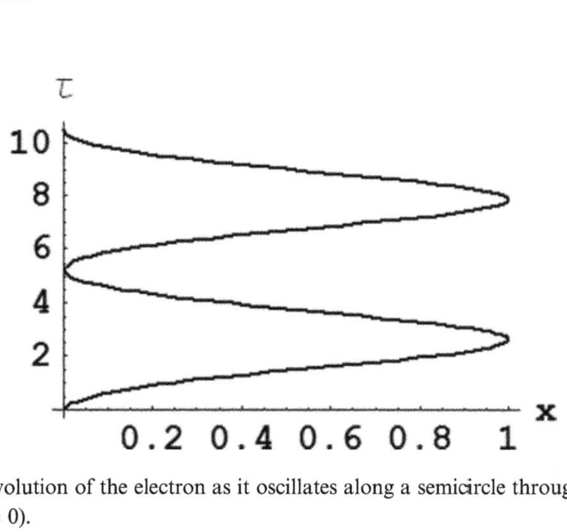

Figure 7.3. The time evolution of the electron as it oscillates along a semicircle through the north pole of the upper hemisphere ($K = 0$).

$$T = 10.488\,[mr^4/(2eD)]^{1/2}. \tag{7.21}$$

Then the author of paper [9] returned to the general situation where the quantity K takes an arbitrary value from the interval (0, $K_{max} = 2/3^{3/2}$). From equation (7.13) one obtains the following general dependence of τ on x (i.e., of τ on cosθ)

$$\tau(x,\ K) = \pm\int_{x_2(K)}^{x} dz/(-z^3 + z - K)^{1/2}. \tag{7.22}$$

In equation (7.22), the lower limit of the integration is the smaller of the two turning points. Figure 7.4 presents the evolution of the scaled time τ during one period of the θ-motion for the following three values of the quantity K: $K = 0.1$ (solid line), $K = 0.2$ (dashed line), and $K = 0.3$ (dotted line). One period of the θ-motion corresponds to x varying from $x_2(K)$ to $x_3(K)$ and back to $x_2(K)$.

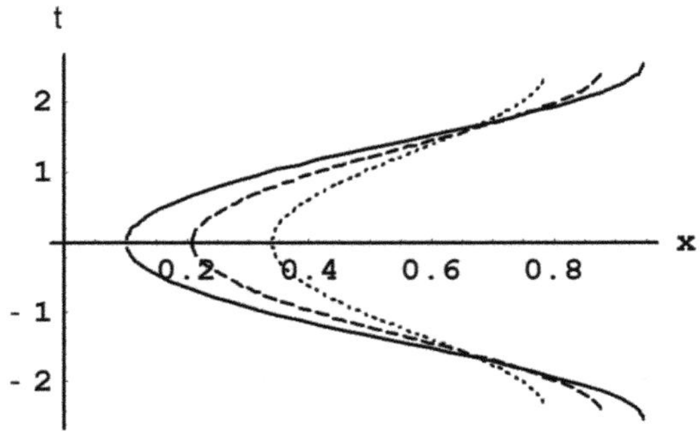

Figure 7.4. Evolution of the scaled time τ during one period of the θ-motion for $K = 0.1$ (solid line), $K = 0.2$ (dashed line), and $K = 0.3$ (dotted line). Reprinted from [9] with permission from MDPI.

The author of paper [9] calculated analytically the integral in equation (7.22) for $K \ll 1$- by expanding it in the Taylor series up to the terms linear in K. The result was as follows:

$$\tau(x,\ K) \approx \pm\ \{f(x) - f[x_2\,(K)] + Kg(x) - Kg[x_2(K)]\} \tag{7.23}$$

where

$$f(x) = -\ 2i\mathbf{F}[\arcsin\,(-x)^{1/2},\ -1] \tag{7.24}$$

$$g(x) = [3x^2 - 2 - x^2\,(1 - x^2)^{1/2}\ {}_2F_1(3/4,\ 1/2,\ 7/4,\ x^2)]/[2(x - x^3)^{1/2}]. \tag{7.25}$$

In equation (7.25), ${}_2F_1(a, b, c, z)$ is the hypergeometric function.

Figure 7.5 demonstrates how accurate the approximate analytical result is for $\tau(x, K)$ from equation (7.23) for $K = 0.01$ (solid line), where it is compared with the corresponding exact result of the numerical integration in equation (7.22) (dashed line). Figure 7.5 shows that the analytical result is very accurate.

Then in paper [9] there has been calculated the scaled period T_θ of the θ-motion. The calculation yielded the following result (in units of $mr^4/(2eD)$):

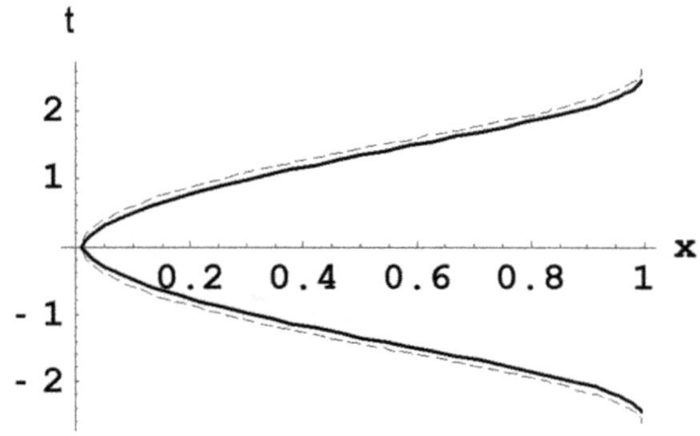

Figure 7.5. Comparison of the approximate analytical result for $\tau(x, K)$ from equation (4.23) for $K = 0.01$ (solid line) with the corresponding exact result obtained by the numerical integration in equation (7.22) (dashed line). Reprinted from [9] with permission from MDPI.

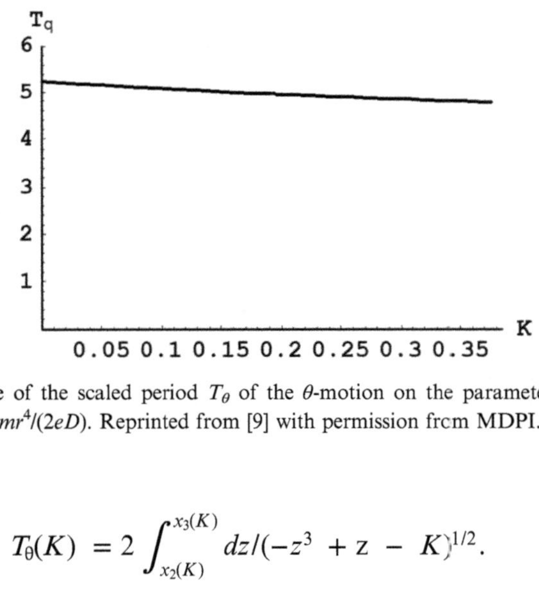

Figure 7.6. Dependence of the scaled period T_θ of the θ-motion on the parameter $K = p_\varphi^2/(2meD)$. The period T_θ is in units of $mr^4/(2eD)$. Reprinted from [9] with permission from MDPI.

$$T_\theta(K) = 2\int_{x_2(K)}^{x_3(K)} dz/(-z^3 + z - K)^{1/2}. \quad (7.26)$$

Figure 7.6 displays the dependence $T_\theta(K)$.

An explicit formula for the scaled period of the θ-motion has been obtained in paper [9] for $K \ll 1$:

$$T_\theta(K) \approx 2\{f(1 - K/2) - f(K) + Kg(1 - K/2) - Kg(K)\}. \quad (7.27)$$

In equation (4.27), functions f and g are defined by equations (4.24) and (4.25), respectively.

Then the author of paper [9] continued by the analysis of the φ-motion. Equation (4.5) can be reformulated as follows:

$$d\varphi = p_\varphi dt/[mr^2(1 - x^2)]. \quad (7.28)$$

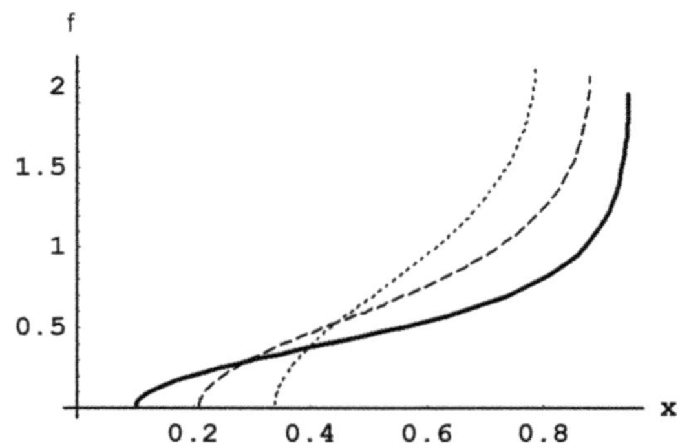

Figure 7.7. Dependence of φ on $x = \cos\theta$ during one half-period of the θ-motion for $K = 0.1$ (solid line), $K = 0.2$ (dashed line), and $K = 0.3$ (dotted line). Reprinted from [9] with permission from MDPI.

Then he utilized equation (7.3) as the relation between dt and dx and integrated with respect to x. As a result, he found the following relation between the angular variable φ and the angular variable $x = \cos\theta$:

$$\varphi(K, x) = K^{1/2}\int_{x_2(K)}^{x} dz/[(1 - z^2)(-z^3 + z - K)^{1/2}]. \tag{7.29}$$

Figure 7.7 provides the visualization of the function $\varphi(x)$ as x changes from $x_2(K)$ to $x_3(K)$ (i.e., for one half-period of the θ-motion) for three different values of K: $K = 0.1$ (solid line), $K = 0.2$ (dashed line), and $K = 0.3$ (dotted line). From figure 7.7 it is clear that with the growth of the value of K, the angle φ depends on x more and more strongly; also over one half-period of the θ-motion the variation of φ becomes slightly larger.

In paper [9] there was derived the following explicit formula for $\varphi(K, x)$ for the case where $K \ll 1$:

$$\varphi(K, x) » K^{1/2}[j(x) - j(K)]. \tag{7.30}$$

In equation (7.30), $j(x)$ stands for

$$j(x) = [x/(1 - x^2)]^{1/2}\left\{1 - (x - 1/x)^{1/2}\,\mathbf{F}[\text{arccsc}(x^{1/2}), -1]\right\}. \tag{7.31}$$

In equation (7.31), $\mathbf{F}(\alpha, q)$ is the incomplete elliptic integral of the first kind. For deriving equation (7.30), the author of paper [9] employed the fact that $x_2(K) \approx K$ for $K \ll 1$.

Figure 7.8 demonstrates, by the example of the subcase of $K = 0.02$, how accurate the approximate explicit formula is for $\varphi(K, x)$ from equation (7.30) (solid line) in comparison to the corresponding exact result from equation (7.29) (dashed line). Figure 7.8 shows that the approximate explicit result is very accurate.

The variation of the angle φ during one half-period of the θ-motion is given by the following expression:

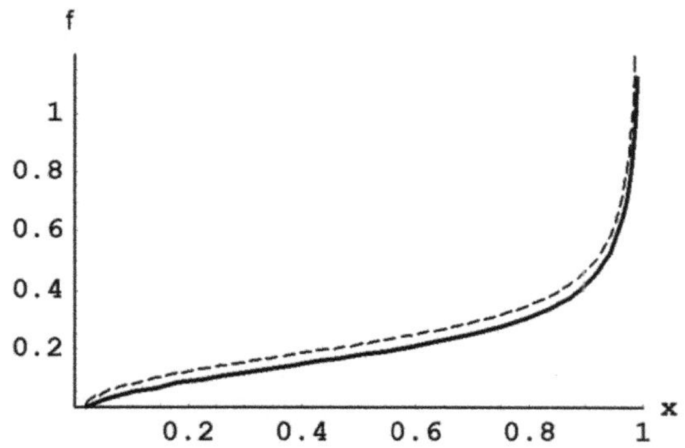

Figure 7.8. Comparison of the approximate analytical result for $\varphi(K, x)$ from equation (7.30) for $K = 0.02$ (solid line) with the corresponding exact result from equation (7.29) (dashed line). Reprinted from [9] with permission from MDPI.

$$\Delta\varphi(K) = K^{1/2}\int_{x_2(K)}^{x_3(K)} dz/[(1 - z^2)(-z^3 + z - K)^{1/2}]. \tag{7.32}$$

For $K \ll 1$, one has $\Delta\varphi \approx \pi/2$, while for K close to its maximum possible value for the bound motion ($K = K_{max} = 2/3^{3/2}$) one gets $\Delta\varphi \approx \pi/2^{1/2}$, so that

$$1/2 < \Delta\varphi/\pi < 1/2^{1/2}. \tag{7.33}$$

With the increase of the quantity K, the angular change $\Delta\varphi$ monotonically grows.

The combined $\theta\varphi$-motion demonstrates the oscillatory-precessional behavior of the electron: it engages simultaneously in the oscillations in the meridional direction in the upper hemisphere and the precession along parallels of latitude. During one period $T_\theta(K)$ of the θ-oscillation, the angle φ changes by $\Delta\varphi(K)$ presented in equation (7.32). Generally speaking, the $\theta\varphi$-motion is not truly periodic, but is *conditionally-periodic*—because generally, the angular variation $\Delta\varphi(K)$ is not equal to $n\pi/m$ (n and m being small integers). So, generally, the trajectory of the electron is not a closed curve.

However, in some situations, where

$$\Delta\varphi(K) = n\pi/m. \tag{7.34}$$

the electron would follow a close curve: the motion would become periodic. In paper [9] the author gave some specific examples. The chosen example were ($n = 2$, $m = 3$), ($n = 3$, $m = 5$), and ($n = 4$, $m = 7$). Figure 7.9 demonstrates what are the values of K where the curve $\Delta\varphi(K) / \pi$ intersects the rational number n/m for these three pairs of n and m.

Below are given further details for all three examples. In the situation where of $n = 2$ and $m = 3$, the angle φ completes two full circles during three periods of the θ-oscillations. This occurs at $K = 0.2103$.

In the situation where $n = 3$ and $m = 5$, the angle φ completes three full circles during five periods of the θ-oscillations. This occurs at $K = 0.0632$.

In the situation where of $n = 4$ and $m = 7$, the angle φ completes four full circles during seven periods of the θ-oscillation. This occurs at $K = 0.0310$.

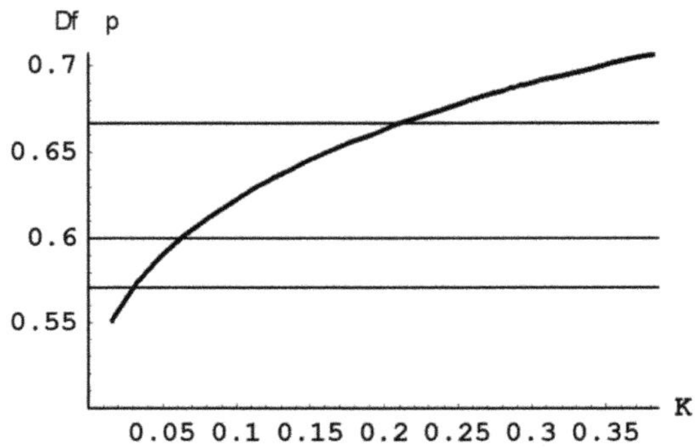

Figure 7.9. Intersections of the curve $\Delta\varphi(K) / \pi$ (thick line) with the tree horizontal lines corresponding to 2/3 (the top thin line), 3/5 (the middle thin line), and 4/7 (the bottom thin line). For the values of K, corresponding to these intersections, the motion, which is generally conditionally-periodic, becomes truly periodic. Reprinted from [9] with permission from MDPI.

These examples show, in particular, that the θ-oscillations can actually occur on the same time scale as the φ-precessions. These two simultaneous motions can occur on the same scale also in the general case of the conditionally-periodic trajectories of the electron. This demonstrates that the author of paper [8] made an incorrect statement while writing that 'in general the motion should consist of large oscillations with respect to the polar angle θ combined with a slow precession about the z-axis'.

In summary, while considering the classical bound motion of a Rydberg electron around a polar molecule, the author of paper [9] derived the analytical relation between the two angular variables θ and φ in the form of a one-fold integral. Also for relatively small values of the projection p_φ of the angular momentum on the axis of the electric dipole ($K = p_\varphi^2 / (2meD) << 1$), he obtained an explicit formula for $\varphi(\theta)$.

For the situation where the electron oscillates along a semicircle crossing the north pole (the case of $K = 0$), in paper [9] there was produced an explicit formula for the dependence of the time t on θ. Then for the opposite scenario where $K = K_{max} = 2/3^{3/2}$ (the electron revolves along the parallel of latitude corresponding to $\theta = 0.9553$ rad = 54.74 degrees), there was derived an explicit formula for the period of the revolution.

In addition, the author of paper [9] derived the time evolution $\theta(t)$ and the period T_θ of the θ-oscillations: in the general case—in the form of a one-fold integral, and in the particular case of $K \ll 1$—explicitly. He also derived the change of the angle φ during one half-period of the θ-oscillations in the general case—in the form of a one-fold integral.

In paper [9] there was also derived a condition, under which the motion of the electron would be periodic (the trajectory would be a closed curve), rather than conditionally-periodic.

Finally, it was demonstrated that in this oscillatory-precessional motion of the electron, both the θ-oscillations and the φ-precessions can take place on the same time scale.

Thus, the Rydberg electron around a polar molecule exhibits a complicated dynamics and the above analytical results offer a physical insight into it.

References

[1] Connolly K and Griffiths D J 2007 *Am. J. Phys.* **75** 524
[2] Fermi E and Teller E 1947 *Phys. Rev.* **72** 399
[3] Turner J E and Fox K 1968 *J. Phys.* A **1** 118
[4] Kryukov N and Oks E 2013 *Can. J. Phys.* **91** 715
[5] Kryukov N and Oks E 2012 *Phys. Rev.* A **85** 054503
[6] Fox K 1968 *J. Phys.* A **1** 124
[7] Jones R S 1995 *Am. J. Phys.* **63** 1042
[8] McDonald K T 1996 http://hep.princeton.edu/~mcdonald/examples/dipole.pdf
[9] Oks E 2020 *Symmetry* **12** 1275

IOP Publishing

Advances in Physics of Rydberg Atoms and Molecules

Eugene Oks

Chapter 8

The possibility of measuring nuclear shapes by using spectral lines of muonic hydrogenic ions

Some properties of nuclei can be uncovered in atomic experiments—see, e.g., review [1] and references therein. For instance, nuclear sizes have been determined from the experimental values of isotope shifts by assuming the spherical shape of the nuclei and the uniform charge distribution—see, e.g., book [2]. The corresponding results of the Gaussian and exponential charge distribution are contained in paper [3]. The Lamb shift has been also calculated assuming the spherical shape of the nuclei—see, e.g., paper [4].

On the other hand, non-spherical nuclear shapes, including ellipsoid-type shape (oblate or prolate) are known to exist for a variety of nuclei—see, e.g., papers [5–7] and references therein. In paper [8], which we follow here, the author considered ellipsoidal nuclei, having the charge distributed uniformly, and derived the spectral line shift for the corresponding hydrogenic ions.

He wrote:

> In this situation the contribution to the shift originates not only from the part of the ionic ('atomic') wave function inside the nucleus, as was the case for spherical nuclei, but also from the modification of the ionic wave function outside the nucleus—in distinction to the case of spherical nuclei. Here and below by ionic wave function we mean the wave function of the bound charged lepton, e.g., electron or muon.

In paper [8] it was demonstrated that after taking into account the ellipsoid-like shape of the nuclei, the spectral line shift for *muonic* ions can differ very significantly (by several times) from the corresponding shift for nuclei of the spherical shape. This can be the basis of a new method for measuring the quadrupole moment of the nuclei and the quantity related to it called beta-parameter. Below are some details.

doi:10.1088/978-0-7503-3939-1ch8

The author of paper [8] analyzed the following system: a hydrogenic ion, having an oblate or prolate nucleus (the nuclear charge being Z) and either electron or muon. In paper [8] there were utilized atomic units $\hbar = e = m_e = 1$. The size of the ellipsoid was considered to be $2a$ in the plane of equator and $2c$ along the axis of the ellipsoid.

In the literature, outside the nucleus, the potential was typically written in the following approximate form:

$$U_{\text{outside}}(r, \cos\theta) = -(Z/r)[1 + \beta(a/r)^2 P_2(\cos\theta)]. \tag{8.1}$$

In equation (8.1)

$$P_2(\cos\theta) = (3\cos^2\theta - 1)/2 \tag{8.2}$$

is one of Legendre polynomials. The coefficient β, entering equation (8.1), is connected to the ellipsoid quadrupole moment D by the following formula

$$\beta = D/(Za^2) \tag{8.3}$$

with

$$D = 2Z(c^2 - a^2)/5 \tag{8.4}$$

(see, e.g., book [9]).

The potential from equation (8.1) is encountered also in the description of other physical systems. One example is a hydrogenlike atom/ion (in the model of a point-like nucleus) subjected to high-frequency laser radiation—see papers [10, 11]. Another example is the motion of a satellite in the field of the oblate Earth. The most important physical feature of all these systems, described by the potential of the type (8.1) is the higher than geometrical (i.e., algebraic) symmetry, leading to the conservation of the square of the angular momentum, while from the geometrical (axial) symmetry followed only the conservation of the projection of the angular momentum on the axis of the symmetry.

For nonzero the angular momentum quantum number L, the first nonvanishing correction to the energy has the form (see, paper [10]):

$$E^{(2)}_{nLM} = (2\beta a^2 Z^4 \mu^3/n^3)[3M^2 - L(L+1)] \\ /[L(L+1)(2L-1)(2L+1)(2L+3)]. \tag{8.5}$$

In equation (8.5), μ is the reduced mass either of the pair 'nucleus–electron' (in which case μ is close to unity) or of the pair of the pair 'nucleus–muon' (in which case μ is about 200)

The corresponding result for $L = 0$ is [10]:

$$E^{(2)}_{n00} = 2\beta a^2 Z^4 \mu^3/(3n^3). \tag{8.6}$$

The averaging of $E^{(2)}{}_{nLM}$ from equation (8.5) over the M-sublevels yields zero. Thus, after the averaging, there remains only $E^{(2)}{}_{n00}$ presented in equation (8.6). By utilizing equations (8.3) and (8.4), one can rewrite equation (8.6), which is the energy correction (shift) caused by the non-sphericity of the potential *outside* the nucleus (the correction related only the energy levels of $L = 0$), in the form

$$\Delta E_{\text{outside}} = E^{(2)}_{n00} = 4(c^2 - a^2)Z^4\mu^3/(15n^3). \quad (8.7)$$

Then the author of paper [8] calculated the energy correction (shift) caused by the non-sphericity of the potential *inside* the nucleus. If the nucleus of the radius R would be spherical, then the potential inside it would be (see, e.g., book [2])

$$U_{\text{inside}}{}^{\text{sphere}}(r) = (Z/R)[r^2/(2R^2) - 3/2], \quad (8.8)$$

so that the perturbation

$$\Delta U_{\text{inside}}{}^{\text{sphere}}(r) = U_{\text{inside}}{}^{\text{sphere}}(r) - (-Z/r), \quad (8.9)$$

yields the following energy shift (for the levels of $L = 0$):

$$\Delta E_{\text{inside}}{}^{\text{sphere}} = 2Z^4\mu^3R^2/(5n^3), \quad (8.10)$$

(the corresponding shift was zero for $L > 0$ [2]).

For the ellipsoidal nucleus, having the same volume V as the sphere of the radius R, the potential inside it is:

$$U_{\text{inside}}(r) = (Z/R)[\rho^2/(2a^2) + z^2/(2c^2) - 3/2], \quad (8.11)$$

In equation (8.11), ρ and z are the cylindrical coordinates, for which the ellipsoid symmetry axis is chosen as the z-axis), and

$$R = [3V/(4\pi)]^{1/3} = a^{2/3}c^{1/3}. \quad (8.12)$$

The representation of equation (8.11) in the spherical polar coordinates is:

$$U_{\text{inside}}(r) = (Z/R)[r^2 \sin^2\theta/(2a^2) + r^2 \cos^2\theta/(2c^2) - 3/2]. \quad (8.13)$$

In equation (8.13), θ is the polar angle.

The perturbation is

$$\Delta U_{\text{inside}}(r) = U_{\text{inside}}(r) - (-Z/r). \quad (8.14)$$

Its average values yields the shift of the energy (non-vanishing only for $L = 0$):

$$\Delta E_{\text{inside}} = [2Z^4\mu^3R^4/(15n^3)](2/a^2 + 1/c^2). \quad (8.15)$$

By combining equations (8.7) and (8.15), the author of paper [8] found the total shift (for $L = 0$ levels) to be as follows:

$$\Delta E_{\text{tot}} = [2Z^4\mu^3R^2/(15n^3)][2(c/a)^{2/3} - 2(a/c)^{2/3} + 2(c/a)^{4/3} + 2(a/c)^{4/3}], \quad (8.16)$$

(we recall that R was given in equation (8.12).

The ratio of the shift for the ellipsoidal nucleus (equation (8.16)) to the shift for the spherical nucleus (equation (8.10)) is:

$$\text{ratio} = \Delta E_{\text{tot}}/\Delta E_{\text{inside}}{}^{\text{sphere}} = [2(c/a)^{2/3} - 2(a/c)^{2/3} + 2(c/a)^{4/3} + 2(a/c)^{4/3}]/3. \quad (8.17)$$

The relation of the ellipticity parameter c/a to the the parameter β, entering equation (8.1), can be found on substituting equation (8.4) in equation (8.3)

$$c/a = (1 + 5\beta/2)^{1/2}. \tag{8.18}$$

The dependence of the ratio = $\Delta E_{tot}/\Delta E_{inside}{}^{sphere}$ on the parameter β was found after substituting equation (8.18) in equation (8.17):

$$\text{ratio} = \Delta E_{tot}/\Delta E_{inside}{}^{sphere} = [2(1 + 5\beta/2)^{1/3} - 2(1 + 5\beta/2)^{-1/3} + 2(1 + 5\beta/2)^{2/3} + (1 + 5\beta/2)^{-2/3}]/3. \tag{8.19}$$

Figure 8.1 presents the above the above ratio versus the parameter β.

Figure 8.2 displays the magnified part of figure 8.1 around the minimum of the ratio equal to 0.454. It corresponds to $\beta = -0.322$.

From figures 8.1 and 8.2 one can see that after taking into account the ellipsoidal shape of the nucleus, the energy shift for muonic ions can indeed change very significantly (by several times). This modified shift can be detected by experimental measurements of the shift of the spectral lines of the Lyman series for muonic ions.

The author of paper [8] wrote (we updated the numeration of the equations):

'This energy shift and its change compared to the spherical nucleus should be possible to detect by observing the shift of the Lyman spectral lines of muonic ions. It is important to note that since the spectral lines of the Lyman series correspond to the radiative transitions from the upper level nP to the lower (ground) level $1S$ and that only the lower level energy gets shifted by ΔE_{tot} presented in equation (8.16), then the shift S of the Lyman lines *in the frequency scale* is

$$S = -\Delta E_{tot}(n = 1) = -(2Z^4\mu^3R^2/15)[2(1 + 5\beta/2)^{1/3} - 2(1 + 5\beta/2)^{-1/3} + 2(1 + 5\beta/2)^{2/3} + (1 + 5\beta/2)^{-2/3}], \tag{8.20}$$

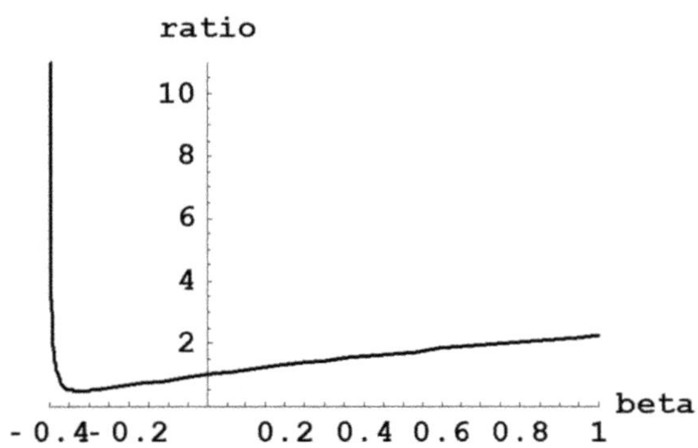

Figure 8.1. Ratio of the energy shift due to the ellipsoid-shaped nucleus to the energy shift due to the corresponding spherical nucleus versus the parameter β, characterizing the degree of the non-sphericity of the nucleus. Reprinted from [8] with permission from MDPI.

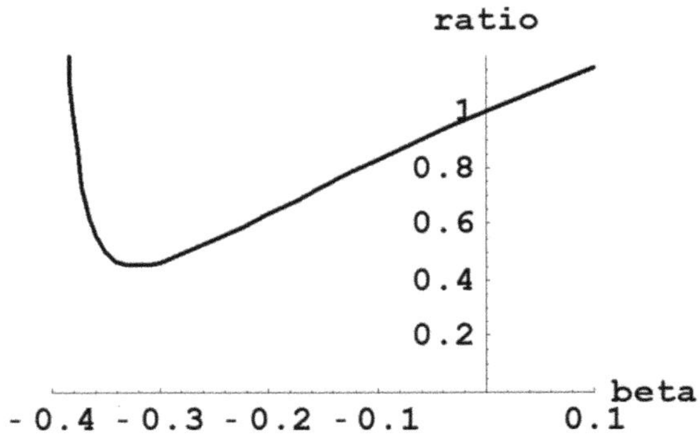

Figure 8.2. The magnified part of the dependence from figure 8.1 around the minimum of this dependence, which is equal to 0.454 and corresponds to $\beta = -0.322$. Reprinted from [8] with permission from MDPI.

where R was defined in equation (8.12). Since $\Delta E_{tot}(n = 1)$ is positive for any β (see figure 8.1), then S is always negative for any β, so that the shift of the Lyman lines *in the wavelength scale* is *always red* for any β.'

A practical recipe from the above results is the following. First, one should measure the experimental shift S of the Lyman lines of a muonic ion. Then, one can utilize equation (8.20) or figure 8.1 and the known nuclear volume V (or the average nuclear radius $R = [3V/(4\pi)]^{1/3}$) for finding the experimental value of the parameter β. Finally, one can determine the experimental value of the nuclear quadrupole moment D by employing the following relation (which is a different representation of equation (8.4)):

$$D = (2ZR^2/5)[(c/a)^{4/3} - (a/c)^{2/3}] = (2ZR^2/5)[(1 + 5\beta/2)^{2/3} - (1 + 5\beta/2)^{-1/3}]. \quad (8.21)$$

A practical formula for the coefficient (in the frequency unit of s $^{-1}$) in front of the square brackets in equation (8.20) is:

$$2Z^4\mu^3R^2/15 = 4.2 \times 10^{-4}Z^4\,[R(\text{fermi})]^2. \quad (8.22)$$

The above analytical results are valid if

$$R_0/r_0 = 5.5 \times 10^{-3}ZA^{1/3} < 1. \quad (8.23)$$

In equation (8.23), the nuclear size is related to the atomic number A as $R_0(\text{a.u.}) = 2.6 \times 10^{-5}\,A^{1/3}$; $r_0(\text{a.u.}) = 1/(\mu Z)$ is the characteristic size of the muonic cloud in the ground state. Usually A = (2–3)Z, so that equation (8.23) can be reformulated in the form:

$$R_0/r_0 = (6.9 - 7.9) \times 10^{-3}Z^{4/3} < 1. \quad (8.24)$$

It is seen that the condition (8.24) is met for $Z < Z_{max} \sim 40$.

Then the author of paper [8] compared the above analytical results to competing effects and provided numerical examples.

He wrote: Since the energy shift due to ellipsoid-shape nucleus can be of the order or several times greater than the energy shift due to the spherical nucleus $\Delta E_{inside}{}^{sphere}$ (as demonstrated in section 2, $\Delta E_{inside}{}^{sphere}$ being presented in equation (10)), it is sufficient to show that the energy shift due to the spherical nucleus significantly exceeds the competing effects.

For the fine structure splitting, which is one of the competing effects, one has:

$$\begin{gathered}\Delta E_{fs}(n) = E(n, j = 3/2) - E(n, j = 1/2) = \alpha^2 \mu Z^4/(4n^3), \ (n > 1); \\ \Delta E_{fs}(1) = 0.\end{gathered} \tag{8.25}$$

In equation (8.25), α is the fine structure constant ($1/\alpha \sim 137$) and j is the quantum number of the total angular momentum. However, the Lyman lines shift, caused by the nuclear shape, is due to the shift of the ground level ($n = 1$) and that the fine structure splitting is zero for the ground level.

Another competing effect is the natural broadening of spectral lines manifesting by the natural width γ, which isproportional to μZ^4. From the corresponding NIST data scaled to muonic ions, the author of paper [8] we obtained the following information

For the line Ly-alpha:

$$\gamma(\text{a. u.}) = 2.4 \times 10^{-6} Z^4, \ \ \Delta E_{inside}{}^{sphere}(n = 2)/\gamma = 1.3 \times 10 A^{2/3}. \tag{8.26}$$

It is seen that for any A, the ratio $\Delta E_{inside}{}^{sphere}(n = 2)/\gamma$ significantly exceeds unity.

For the line Ly-beta:

$$\gamma(\text{a. u.}) = 2.8 \times 10^{-7} Z^4, \ \ \Delta E_{inside}{}^{sphere}(n = 3)/\gamma = 3.2 \times 10^2 A^{2/3}. \tag{8.27}$$

Again, it is seen that for any A, the ratio $\Delta E_{inside}{}^{sphere}(n = 3)/\gamma$ significantly exceeds unity.

For the line Ly-gamma:

$$\gamma(\text{a. u.}) = 6.4 \times 10^{-8} Z^4, \ \ \ \Delta E_{inside}{}^{sphere}(n = 4)/\gamma = 6.0 \times 10^2 A^{2/3}. \tag{8.28}$$

Once again, it is seen that for any A, the ratio $\Delta E_{inside}{}^{sphere}(n = 3)/\gamma$ significantly exceeds unity. So, the Lyman lines shift due to the nuclear shape is greater than the natural width for any nucleus by many orders of magnitude.

Then the author of paper [8] compared the Lyman lines shift due to the shape of the nuclei to the unperturbed frequency $\omega_0(n)$, which in atomic units is

$$\omega_0(n) = (\mu Z^2/2)(1 - 1/n^2). \tag{8.29}$$

The idea behind the comparison was to ensure that the typical experimental spectral resolution would be sufficient for observing this shift. Since the shift for the ellipsoidal nucleus is of the order or several times greater than the shift for the spherical nucleus, it was enough to make sure that the latter can be resolved experimentally.

For the case of the spherical nucleus ($c = a$), for muonic ions, the shift of the frequency of the Lyman lines is (see equation (8.20)):

$$S_{\mathrm{spher}} = -2Z^4\mu^3R^2/5. \tag{8.30}$$

Then for the ratio under consideration one gets:

$$S_{\mathrm{spher}}/\omega_0(n) = -4Z^2\mu^2R^2/[5(1 - 1/n^2)] = -2.4 \times 10^{-5}Z^2A^{2/3}. \tag{8.31}$$

Therefore, the corresponding estimate for ellipsoidal nuclei is:

$$S/\omega_0(n) > S_{\min}/\omega_0(n) \sim -2.4 \times 10^{-5}Z^2A^{2/3}. \tag{8.32}$$

Then the author of paper [8] wrote:

'Below we present numerical values of $S_{\min}/\omega_0(n)$ for the muonic Lyman lines for several nuclei that are listed, e.g., in paper [5] as known to be oblate ($\beta < 0$).

${}^{20}{}_{6}C_{14}$ *(oblate in the nuclear ground state)*:

$S_{\min}/\omega_0(2)$ ~−0.008 (for Ly-α), $S_{\min}/\omega_0(3)$ ~−0.007 (for Ly-β), $S_{\min}/\omega_0(4)$ ~−0.007 (for Ly-γ)

${}^{42}{}_{14}Si_{28}$ *(oblate in the nuclear ground state)*:

$S_{\min}/\omega_0(2)$ ~−0.07 (for Ly-α), $S_{\min}/\omega_0(3)$ ~−0.06 (for Ly-β), $S_{\min}/\omega_0(4)$ ~−0.06 (for Ly-γ)

${}^{12}{}_{6}C_{6}$ *(oblate in the 1st excited 2^+ nuclear state)*:

$S_{\min}/\omega_0(2)$ ~−0.006 (for Ly-α), $S_{\min}/\omega_0(3)$ ~−0.005 (for Ly-β), $S_{\min}/\omega_0(4)$ ~−0.005 (for Ly-γ)

${}^{28}{}_{14}Si_{14}$ *(oblate in the 1st excited 2^+ nuclear state)*:

$S_{\min}/\omega_0(2)$ ~−0.06 (for Ly-α), $S_{\min}/\omega_0(3)$ ~−0.05 (for Ly-β), $S_{\min}/\omega_0(4)$ ~−0.05 (for Ly-γ)

${}^{34}{}_{16}S_{18}$ *(oblate in the 1st excited 2^+ nuclear state)*:

$S_{\min}/\omega_0(2)$ ~−0.08 (for Ly-α), $S_{\min}/\omega_0(3)$ ~−0.07 (for Ly-β), $S_{\min}/\omega_0(4)$ ~−0.07 (for Ly-γ)

${}^{36}{}_{18}Ar_{18}$ *(oblate in the 1st excited 2^+ nuclear state)*:

$S_{\min}/\omega_0(2)$ ~−0.11 (for Ly-α), $S_{\min}/\omega_0(3)$ ~−0.09 (for Ly-β), $S_{\min}/\omega_0(4)$ ~−0.09 (for Ly-γ)

Thus, for all of the above examples, the relative shift of the muonic Lyman lines caused by ellipsoid-shaped nuclei is sufficiently large to be detected by spectrometers.'

In summary, the most important results presented in paper [8] were that after taking into account the ellipsoidal shape of the nuclei, the shift of muonic ions spectral lines can differ by several times from the one for spherical nuclei, and that this shift is much greater than the competing effects, and that spectrometers can detect it. Therefore, the results obtained in paper [8] offer a new method for measuring the quadrupole moment of nuclei and of the standard parameter β.

References

[1] Neugart R 1992 *Nuclear Shapes and Nuclear Structures at Low Excitation Energies* ed M Vergnes, J Sauvage, P-H Heenen and H Tuan Duong (New York: Plenum Press) p 59

[2] Flügge S 1974 *Practical Quantum Mechanics* Part II (Berlin: Springer) p 191
[3] Friar J L 1979 *Ann. Phys.* **122** 151
[4] Krutov A A, Martynenko A P, Martynenko F A and Sukhorukova O S 2016 *Phys. Rev.* A **94** 062505
[5] Hamamoto I 2014 *Phys. Rev.* C **89** 057301
[6] Podolyak Z 2012 *J. Phys. Conf. Series* **381** 012052
[7] Lucas R 2001 *Europhysics News* p 5
[8] Oks E 2018 *Atoms* **6** 14
[9] Landau L D and Lifshitz E M 1987 *Classical Theory of Fields* section 41 (Oxford: Butterworth-Heinemann)
[10] Gavrilenko V P, Oks E and Radchik A V 1985 *Opt. Spectrosc.* **59** 411
[11] Nadezhdin B B and Oks E 1986 *Sov. Tech. Phys. Lett.* **12** 512
[12] Beletsky V V 2012 *Essays on the Motion of Celestial Bodies* (Basel: Birkhäuser-Springer) ch 2

IOP Publishing

Advances in Physics of Rydberg Atoms and Molecules

Eugene Oks

Chapter 9

Muonic-electronic negative hydrogen ion: circular states

Investigations of muonic atoms and molecules are important for several reasons, such as muon-catalyzed fusion (see, e.g., [1–3] and references therein), laser-control of nuclear processes [4], the hunt for strongly interacting massive particles, once hypothesized as a possible dark matter candidate (see, e.g., [5] and references therein)[1].

As for investigations of the negative ion of hydrogen H^- (i.e., the epe-system), they are important because it is one of the simplest atomic systems exhibiting strong correlations, and because it plays a significant role in the D-layer of the Earth's atmosphere, in the atmosphere of stars (including the Sun), and in the design of particle accelerators [10].

In paper [11], which we overview here, the authors conjoined the above two research areas, by considering the μpe-system: the muonic-electronic negative hydrogen ion. They focused on the circular states. They demonstrated that the electronic motion can be considered as the slow subsystem, while the muonic motion is the rapid subsystem.

The authors of paper [11] showed that this physical system can behave as a quasimolecule where the muon rapidly rotates about the axis connecting the electron and the proton, while this axis slowly rotates. They also demonstrated that the muonic-emitted spectral lines undergo a red shift, which can be used for detecting such a system. Below are some details.

[1] By now the most probable candidate for dark matter or at least a part of it seems to be the second flavor of hydrogen atoms (SFHA)—see papers [6, 7]. This is because it is the only explanation of dark matter that has the following four features simultaneously: (1) it has the experimental confirmation, namely from the analysis of atomic experiments [8]; (2) it does not go beyond the Standard Model (and thus is favored by the Occam razor principle); (3) it is based on the standard quantum mechanics, namely on the Dirac equation—without any change to the physical laws (and thus again is favored by the Occam razor principle); (4) it explains the recent puzzling astrophysical observations by Bowman *et al* [9].

The authors of paper [11] analyzed the configuration depicted in figure 9.1. They used the term 'internuclear axis' for the line connecting the electron with the proton. Throughout their paper, they utilized atomic units.

The authors of paper [11] proceeded from the results of papers [12, 13]. The Hamiltonian of the system in paper [11] was

$$H = (p_z^2 + p_\rho^2 + p_\varphi^2/\rho^2)/(2m) - Z/(z^2 + \rho^2)^{1/2} - Z'/[(R - z)^2 + \rho^2]^{1/2}. \quad (9.1)$$

In equation (9.1), R is the separation between the 'nuclei', Z and Z' are the charges of the effective nuclei (i.e., $Z = 1$ and $Z' = -1$), and $m = 206.768\,2746$ a.u. is the muonic mass. The authors used the cylindrical coordinates (ρ, φ, z), such that the charge Z was at the origin and the charge Z' is at $z = R$.

The momentum $p_\varphi|$ is conserved because the coordinate φ is a cyclic:

$$|p_\varphi| = \text{const} = L. \quad (9.2)$$

The Hamiltonian for the ρ- and z-motions is

$$H_{z\rho} = (p_z^2 + p_\rho^2)/2 + U_{\text{eff}}(z, \rho). \quad (9.3)$$

In equation (9.3), $U_{\text{eff}}(z, \rho)$ is effective potential energy:

$$U_{\text{eff}}(z, \rho) = L^2/(2m\rho^2) - Z/(z^2 + \rho^2)^{1/2} - Z'/[(R - z)^2 + \rho^2]^{1/2} \quad (9.4)$$

which coincides with the total energy ($E(z, \rho) = U_{\text{eff}}(z, \rho)$) because $p_z = p_\rho = 0$ in circular states.

After setting $Z = 1$, $Z' = -1$ and introducing the scaled quantities

$$w = z/R,\ v = \rho/R,\ \varepsilon = -ER,\ \ell = L/(mR)^{1/2},\ r = mR/L^2 \quad (9.5)$$

the authors of paper [11] got the scaled energy ε of the muon in the form:

$$\varepsilon = 1/(w^2 + v^2)^{1/2} - 1/[(1 - w)^2 + v^2]^{1/2} - \ell^2/(2v^2). \quad (9.6)$$

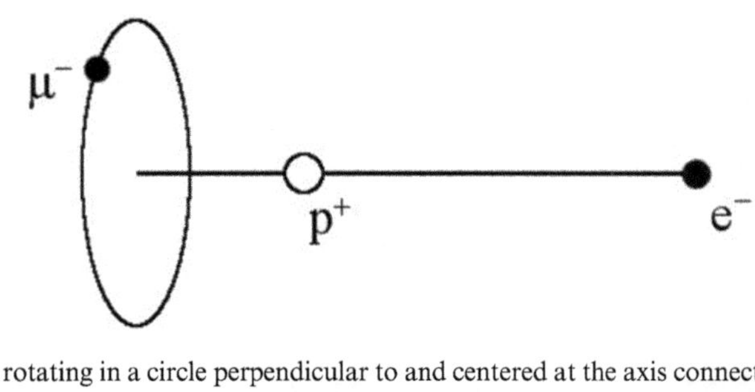

Figure 9.1. A muon rotating in a circle perpendicular to and centered at the axis connecting the proton and the electron. Reprinted from [11] with permission from the Canadian Journal of Physics.

One of the conditions of the equilibrium, namely $\partial\varepsilon/\partial w = 0$, leads to:

$$[(1 - w)^2 + v^2]^{3/2} / (w^2 + v^2)^{3/2} = (w - 1)/w. \tag{9.7}$$

Obviously, there should be $(w - 1)/w > 0$. Therefore, for the allowed ranges of the scaled coordinate w, the authors of paper [11] obtained: $-\infty < w < 0$ and $1 < w < +\infty$. Consequently, between the electron and proton (i.e., in the range of $0 < w < 1$, there is no equilibrium. The explicit expression from the quantity $v^2 = p$, following from equation (9.7), is:

$$p(w) = w^{2/3}(w - 1)^{2/3}[w^{2/3} + (w - 1)^{2/3}]. \tag{9.8}$$

From the other condition of the equilibrium μpe $\rightarrow$ μ + pe follows:

$$\ell^2 = p^2\{1/(w^2 + p)^{3/2} - 1/[(1 - w)^2 + p]^{3/2}\}. \tag{9.9}$$

It is seen that the right side of equation (9.9) should be positive, leading to the requirement $w < 1/2$. Consequently, only in the range of $-\infty < w < 0$ does there exist the equilibrium with respect to both scaled coordinates w and v.

By combining equations (9.5) and (9.9), the authors of paper [11] obtained:

$$r = p - 2\left\{1/(w^2 + p)^{3/2} - 1/[(1 - w)^2 + p]^{3/2}\right\}^{-1}. \tag{9.10}$$

Since the dependence $p(w)$ is provided by equation (9.8), then equation (9.10) gives the dependence of the scaled 'internuclear' distance r on the scaled internuclear coordinate w.

The authors of paper [11] combined equations (9.6), (9.8), and (9.9) and got the dependence $\varepsilon(w)$. Then they used the relations $E = -\varepsilon/R$ and $R = rL^2/m$ for expressing $E = -(m/L^2)\varepsilon_1$ where $\varepsilon_1 = \varepsilon/r$. The combination of $\varepsilon_1(w)$ with $r(w)$ from equation (9.10) represents the parametric dependence $\varepsilon_1(r)$, which are the energy terms.

Then the authors of paper [11] introduced the notation $\gamma = (1 - 1/w)^{1/3}$ to have the dependence $\varepsilon_1(r)$ via the parameter γ, the region $-\infty < w < 0$ being in the correspondence with the region $1 < \gamma < \infty$. Thereafter, the parametric dependence $\varepsilon_1(r)$ simplified to the form:

$$\varepsilon_1(\gamma) = (1 - \gamma)^4(1 + \gamma^2)^2 / [2(1 - \gamma + \gamma^2)^2(1 + \gamma^2 + \gamma^4)] \tag{9.11}$$

$$r(\gamma) = (1 + \gamma^2 + \gamma^4)^{3/2}/[\gamma(1 + \gamma^2)^2]. \tag{9.12}$$

Figure 9.2 illustrates the parametric dependence of the scaled energy $\varepsilon_1 = (L^2/m)E$ on the scaled internuclear distance $r = (m/L^2)R$, i.e., the classical energy terms. In fact, this figure presents two coinciding energy terms. This is the manifestation of the double degeneracy with respect to the sign of the projection of **L** on the proton–electron axis.

The term in figure 9.2 begins at $w = -\infty$, corresponding to $\gamma = 1$. By using equation (9.12), the minimum value of R was found to be

$$R_{\min} = (3^{3/2}/4)L^2/m. \tag{9.13}$$

Since $m = 206.7682746$, then $R = 0.00628258\ L^2$.

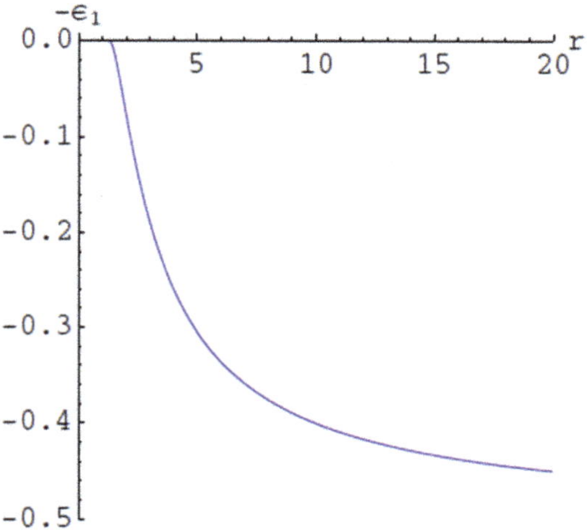

Figure 9.2. Classical energy terms: the scaled energy $-\varepsilon_1 = (L^2/m)E$ versus the scaled internuclear distance $r = (m/L^2)R$. Reprinted from [11] with permission from the Canadian Journal of Physics.

The muon revolves with the following frequency:

$$\Omega = L/(m\rho^2) = L/(mR^2v^2) = L/(mR^2p). \tag{9.14}$$

In view of equation (9.5), it is possible to rewrite equation (9.14) as $\Omega = (m'/L^3)f$ with $f = 1/(pr^2)$. By utilizing equations (9.8) and (9.12), the authors of paper [11] got:

$$\Omega = (m'/L^3)f(\gamma), \ f(\gamma) = (1 + \gamma^2)^3(1 - \gamma^3)^2 / (1 + \gamma^2 + \gamma^4)^3. \tag{9.15}$$

Figure 9.3 depicts the dependence of the scaled revolution frequency of the muon $f = (L^3/m)\omega$ on the scaled internuclear distance $r = (m/L^2)R$.

One can see that the scaled revolution frequency of the muon is practically equal to its maximum value $f_{max} = 1$ for almost any value of r. This means that the revolution frequency Ω of the muon is practically equal to its maximum value

$$\Omega_{max} = m/L^3 \tag{9.16}$$

for almost any value of R.

Then the authors of paper [11] analyzed the stability of the muon motion by considering small oscillations of its ρ and φ coordinates around their equilibrium values. It turned out that the above classical energy terms represent the stable motion.

After unfreezing the slow subsystem, it can be considered as a kind of a 'rigid rotator'. The latter contains (in addition to the proton and the electron) the ring of the uniformly distributed muon charge. The electron potential energy has the form:

$$E_e = M^2/(2R^2) - 1/R + 1/[\rho^2 + (R - z)^2]^{1/2}. \tag{9.17}$$

In equation (9.17), M is the angular momentum of the electron. The equilibrium corresponds to the zero value of the derivative of E_e with respect to R:

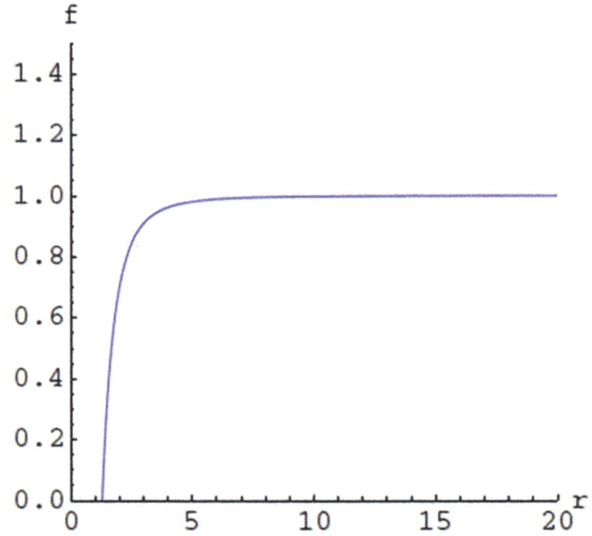

Figure 9.3. The scaled muon revolution frequency $f = (L^3/m)\Omega$ versus the scaled internuclear distance $r = (m/L^2)R$. Reprinted from [11] with permission from the Canadian Journal of Physics.

$$dE_e/dR = -M^2/R^3 + 1/R^2 - (R - z)/[\rho^2 + (R - z)^2]^{3/2} = 0. \tag{9.18}$$

From equation (9.18), the authors of paper [11] obtained the following value of the scaled angular momentum of the electron $\ell_e = M/R^{1/2}$

$$\ell_e^2 = 1 - (1 - w)/[(1 - w)^2 + p]^{3/2}. \tag{9.19}$$

By utilizing equation (9.7), it is possible to rewrite equation (9.19) as follows

$$\ell_e^2 = 1 + w/(w^2 + p)^{3/2} \tag{9.20}$$

or after substituting w by $\gamma = (1 - 1/w)^{1/3}$:

$$\ell_e(\gamma) = [1 - (1 - \gamma)^2(1 + \gamma + \gamma^2)^{1/2}/(1 - \gamma + \gamma^2)^{3/2}]^{1/2}. \tag{9.21}$$

The revolution frequency of the electron is

$$\omega = M/R^2 = \ell_e(\gamma)/R^{3/2}. \tag{9.22}$$

By using equations (9.5) and (9.12), the authors of paper [11] represented the revolution frequency of the electron in the form:

$$\omega = m^{3/2}\ell_e(\gamma)/\{L^3[r(\gamma)]^{3/2}\}. \tag{9.23}$$

The ratio of the revolution frequencies of the muon and the electron is:

$$\Omega/\omega = (1/m^{1/2})f(\gamma)[r(\gamma)]^{3/2}/\ell_e(\gamma). \tag{9.24}$$

We recall that the function $f(\gamma)$ was defined in equation (9.15).

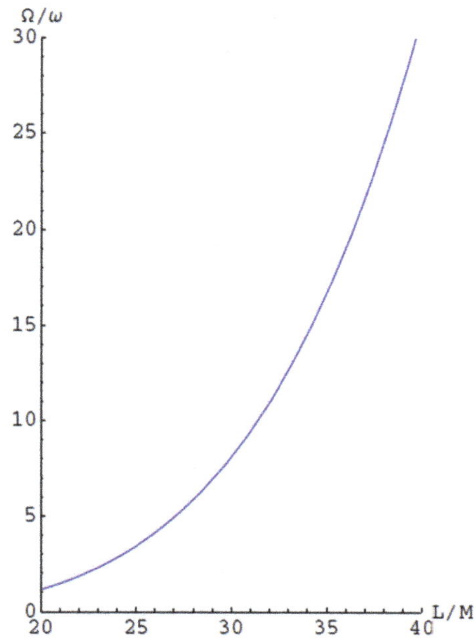

Figure 9.4. The ratio of the muon and electron revolution frequencies Ω/ω versus the ratio of the muon and electron angular momenta L/M. Reprinted from [11] with permission from the Canadian Journal of Physics.

The quantity R can be expressed as either

$$R = L^2 r(\gamma)/m \tag{9.25}$$

or as

$$R = M^2/[\ell_e(\gamma)]^2. \tag{9.26}$$

After equating the right sides of equations (9.25) and (9.26), the authors of paper [11] got the following relation:

$$L/M = m^{1/2}/\{\ell_e(\gamma)[r(\gamma)]^{1/2}\}. \tag{9.27}$$

The duo of equations (9.24) and (9.27) constitute the parametric dependence of the ratio Ω/ω on the ratio L/M, as the parameter γ changes from 1 to ∞. Figure 9.4 depicts this dependence.

From figure 9.4 one can see that the ratio L/M should be noticeably greater than 20 for the frequency ratio Ω/ω to be significantly greater than unity, as the necessary condition for the validity of separating the slow and rapid subsystems.

Another necessary condition of the validity is for the inverse lifetime of the muon $1/T_{\text{life}}$ ($T_{\text{life}} = 2.\ 2\ \mu s = 0.91 \times 10^{11}$ a.u.) to satisfy the following inequality:

$$\Omega T_{\text{life}} >> 1. \tag{9.28}$$

The inequality (9.28) can be rewritten as

$$(m/L^3)T_{\text{life}} >> 1. \tag{9.29}$$

given that the revolution frequency Ω of the muon is practically equal to its maximum value $\Omega_{max} = m/L^3$ for almost any R. From the inequality (9.29), the authors of paper [11] obtained

$$L << L_{max} = (m\, T_{life})^{1/3} = 26600 \tag{9.30}$$

where the value m = 206. 7682746 a.u. has been used. Thus, for any realistic value of of the angular momentum L of the muon, the second validity condition is met.

The muonic hydrogen atom, i.e., the μp-subsystem, would radiate spectral lines at the frequency $\Omega_{max} = m/L^3$. In distinction, the system μpe emits the spectral lines at the frequency Ω that is slightly smaller than Ω_{max}. In other words, in the quasimolecule *μpe*, there is a *red shift* δ of the muon-emitted spectral lines counted from the corresponding spectral lines of a muon in a muonic hydrogen atom:

$$\delta = (\lambda - \lambda_0)/\lambda_0 = (\Omega_{max} - \Omega)/\Omega. \tag{9.31}$$

In equation (9.31), λ_0 and λ are the wavelengths of the spectral lines for the muonic hydrogen atom and the quasimolecule μpe, respectively. By utilizing equation (9.15), the authors of paper [11] rewrote equation (9.31) as follows

$$\delta(\gamma) = 1/f(\gamma) - 1. \tag{9.32}$$

The duo of equations (9.27) and (9.32) constitute the parametric dependence of the relative red shift δ on the ratio of the L/M. Figure 9.5 presents the relative red shift δ versus $L/(m^{1/2} M)$.

The authors of paper [11] wrote:

'In this form the dependence is 'universal', i.e., valid for different values of the mass m: for example, it is valid also for the quasimolecule πpe where there is a pion instead of the muon.'

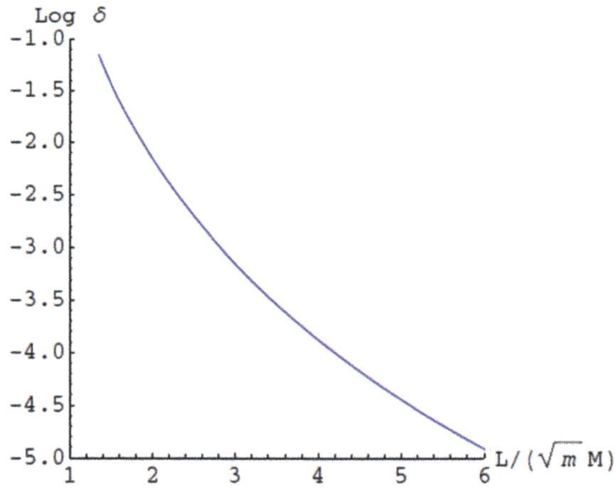

Figure 9.5. Universal dependence of the relative red shift δ of the spectral lines of the quasimolecule μpe (or πpe) on $L/(m^{1/2}M)$, which is the ratio of the muon and electron angular momenta L/M divided by the square root of the mass m of the muon or pion. Reprinted from [11] with permission from the Canadian Journal of Physics.

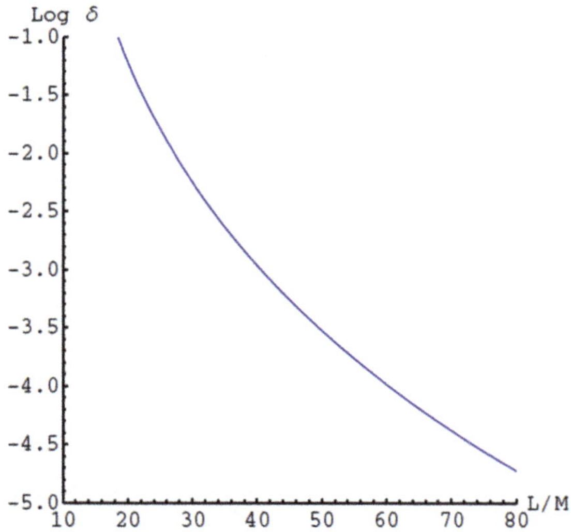

Figure 9.6. Dependence of the relative red shift δ of the spectral lines of the quasimolecule μpe on the ratio of the muon and electron angular momenta L/M. Reprinted from [11] with permission from the Canadian Journal of Physics.

Figure 9.6 presents the relative red shift δ versus L/M, i.e., specifically for the quasimolecule μpe.

From figure 9.6 one can see that for the ratio $L/M < 80$, can be detected by available spectrometers, since their spectral resolution is $\Delta\lambda_{res}/\lambda \sim (10^{-4} - 10^{-5})$. Therefore, the formation of muonic negative hydrogen ions can be detected by observing this red shift.

Figure 9.7 shows the relative red shift δ versus the frequency ratio Ω/ω. One can see with the growth of the frequency ratio Ω/ω, the relative red shift diminishes. Yet, it continues to be within the spectral resolution of available spectrometers.

In summary, the authors of paper [11] analyzed the muonic-electronic negative hydrogen ion by separating rapid (muon) and slow (electron) subsystems. It was counterintuitive that the motion of the heavier lepton was much more rapid than the motion of the lighter lepton.

The authors of paper [11] obtained analytically classical energy terms. They demonstrated that the muonic motion is stable.

The authors of paper [11] also considered relativistic corrections to the motion of the muon (not presented in this chapter) and found them to be $\sim 1/(cL)^2 \sim 0.5 \times 10^{-4}/L^2 << 1$. Thus, these corrections are insignificant.

Finally, the authors of paper [11] focused on the red shift of the spectral lines emitted by the muon in the quasimolecule μpe. They demonstrated that its relative value significantly exceeds the typical spectral resolution of spectrometers, so that it can be observed and serve for the experimental detection of the formation of these quasimolecules.

Finally the authors wrote (we updated here the numeration of the formula and of the references):

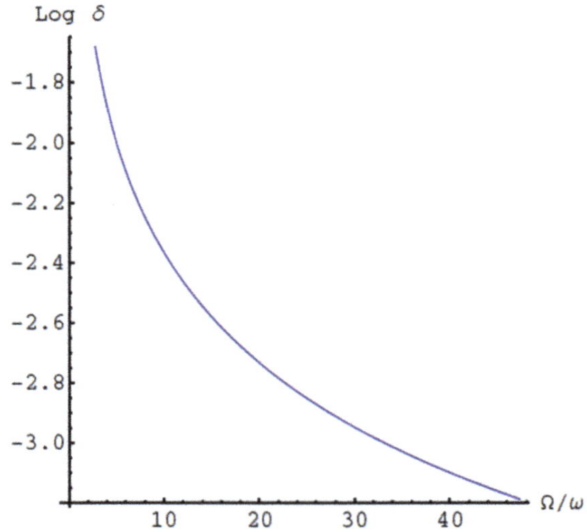

Figure 9.7. Dependence of the relative red shift δ on the ratio of the muon and electron revolution frequencies Ω/ω. Reprinted from [11] with permission from the Canadian Journal of Physics.

'As for physical processes leading to the formation of muonic-electronic negative hydrogen ions, one of the processes could be the following:

$$\text{e} + \mu\text{p} \;\rightarrow\; \mu\text{pe} \tag{9.33}$$

(which sometimes might be followed by the decay $\mu\text{pe} \rightarrow \mu + \text{pe}$). Such formation of the μpe-systems was discussed, e.g., in paper [14], where such systems were called 'resonances'. We also mention that there are methods to create muonic hydrogen atoms μp necessary for the reaction (36) and that these methods lead to the muon being in a highly-excited state (see, e.g., review [15] and paper [16]).'

References

[1] Ponomarev L I 1990 *Contemp. Phys.* **31** 219

[2] Nagamine K 2001 *Hyperfine Interact.* **138** 5

[3] Nagamine K and Ponomarev L I 2003 *Nucl. Phys.* A **721** C863

[4] Chelkowsky C, Bandrauk A D and Corkum P B 2004 *Laser Phys.* **14** 473

[5] Guffin J, Nixon G, Javorsek D II, Colafrancesco S and Fischbach E 2002 *Phys. Rev.* D **66** 123508

[6] Oks E 2020 *E. Res. Astron. Astrophys.* **20** 109

[7] Oks E 2020 *Atoms* **8** 33

[8] Oks E 2001 *J. Phys. B: At. Mol. Opt. Phys.* **34** 2235

[9] Bowman J D, Rogers A E E, Monsalve R A, Mozdzen T J and Mahesh N 2018 *Nature* **555** 67

[10] Rau A R P 1996 *J. Astrophys. Astron.* **17** 113

[11] Kryukov N and Oks E 2013 *Can. J. Phys.* **91** 715

[12] Oks E 2000 *Phys. Rev. Lett.* **85** 2084

[13] Oks E 2000 *J. Phys. B: At. Mol. Opt. Phys.* **33** 3319

[14] Sakimoto K 2008 *Phys. Rev.* A **78** 042509
[15] Measday D F 2001 *Phys. Rep.* **354** 243
[16] Sakimoto K 2010 *Phys. Rev.* A **81** 012511

IOP Publishing

Eugene Oks

Chapter 10

Muonic-electronic quasimolecules based on a fully-stripped multicharged ion

In chapter 9 we overviewed paper [1] where its author analyzed the quasimolecule μpe. Both the muon and the electron were assumed to be in circular states. In chapter 9 it was emphasized that they are important for several reasons, such as muon-catalyzed fusion (see, e.g., [2–4] and references therein), laser-control of nuclear processes [5], the hunt for strongly interacting massive particles, once hypothesized as a possible dark matter candidate (see, e.g., [6] and references therein), though concerning the latter it was noted that the most probable candidate for dark matter or at least a part of it seems to be the second flavor of hydrogen atoms—see papers [7, 8] and some details in chapter 9.

In paper [9], which we overview here, the authors generalized the investigation from paper [1]. They analyzed the quasimolecule μZe, where compared to the study from paper [1] the proton was substituted by a nucleus of charge Z.

The authors of paper [9] demonstrated that for the validity of separating rapid and slow subsystems, the ratio of the angular momenta of the muon and the electron should slightly exceed the corresponding ratio from the case of $Z = 1$. They also showed that for $Z > 1$, the muon binding energy significantly exceeds the corresponding binding energy from the case of $Z = 1$.

The authors of paper [9] also showed that as the nuclear charge Z grows, the red shift of the muon-emitted spectral lines diminishes. However, it continues to be within the spectral resolution of typical spectrometers as a minimum until $Z = 5$, so that this red shift is still possible to observe. Below are some details.

The configuration considered in paper [9] in depicted in figure 10.1.

doi:10.1088/978-0-7503-3939-1ch10

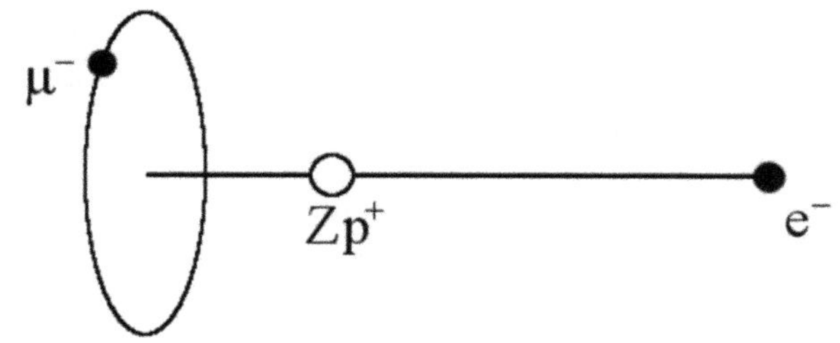

Figure 10.1. A muon rotating in a circle perpendicular to and centered at the axis connecting the nucleus and the electron. Reprinted from [9] with permission from the Canadian Journal of Physics.

10.1 Analytical solution for classical energy terms of the rapid subsystem

In this study we consider a system where a muon rotates in a circle perpendicular to the axis connecting a nucleus of charge Z and an electron—see figure 10.1. Atomic units are used in this article, just as in [1].

In paper [9] the authors investigated the general situation, where $Z > 1$, by employing the same technique as in paper [1]. They called the pair 'nucleus–electron' the 'effective nuclei'. They utilized the atomic units.

The effective potential energy has the form:

$$U_{\text{eff}}(z, \rho) = \frac{L^2}{2m\rho^2} - \frac{Z}{\sqrt{z^2 + \rho^2}} - \frac{Z'}{\sqrt{(R - z)^2 + \rho^2}}. \tag{10.1}$$

Here R is the separation between the effective nuclei; Z and Z' are their charges (specifically, $Z' = -1$); $m = 206.768\,2746$ a.u. is the muonic mass; (ρ, φ, z) are the cylindrical coordinates. The authors of paper [9] introduced the following scaled quantities:

$$w = \frac{z}{R}, \; v = \frac{\rho}{R}, \; \varepsilon = -ER, \; \ell = \frac{L}{\sqrt{mR}}, \; r = \frac{mR}{L^2}. \tag{10.2}$$

Thereafter, for the scaled muonic energy they got:

$$\varepsilon = \frac{Z}{\sqrt{w^2 + p}} - \frac{1}{\sqrt{(1 - w)^2 + p}} - \frac{\ell^2}{2p}. \tag{10.3}$$

In equation (10.3), $p \equiv v^2$ is the squared scaled radial coordinate.

By equating to zero the partial derivatives of ε with respect to w and p, the authors of paper [9] obtained the following two equations:

$$p = w^{2/3}(w - 1)^{2/3}\frac{Z^{2/3}(w - 1)^{4/3} - w^{4/3}}{(w - 1)^{2/3} - Z^{2/3}w^{2/3}} \tag{10.4}$$

$$\ell^2 = p^2\left(\frac{Z}{(w^2 + p)^{3/2}} - \frac{1}{((1 - w)^2 + p)^{3/2}}\right). \tag{10.5}$$

From the fact that the left sides of equations (10.4) and (10.5) are positively defined, the authors of paper [9] deduced the following restriction on the equilibrium values of the scaled coordinate w:

$$-\frac{1}{Z-1} < w < 0. \tag{10.6}$$

By combining equations (10.3)–(10.5), the authors of paper [9] obtained the dependence $\varepsilon(w)$ of the scaled muonic energy on the scaled coordinate w. Thereafter, they utilized the same substitution as in paper [1]:

$$\gamma = \left(1 - \frac{1}{w}\right)^{1/3}. \tag{10.7}$$

Finally, the authors of paper [9] noted that $r = 1/\ell^2$ according to equation (10.2) and obtained the following parametric dependence of the re-scaled muonic energy $\varepsilon_1 = \varepsilon/r$ on r (for any given nuclear charge Z):

$$\varepsilon_1 = \frac{(Z^{2/3}\gamma^4 - 1)^2(Z^{2/3}\gamma(\gamma^3 + 2) - 2\gamma^3 - 1)}{2(\gamma^3 - 1)(\gamma^3 + 1)^3} \tag{10.8}$$

$$r = \frac{\sqrt{(\gamma^6 - 1)^3(\gamma^2 - Z^{2/3})}}{\gamma(Z^{2/3}\gamma^4 - 1)^2}. \tag{10.9}$$

Figure 10.2 exhibits the calculated energy terms for several values of the nuclear charge.

The authors of paper [9] wrote:

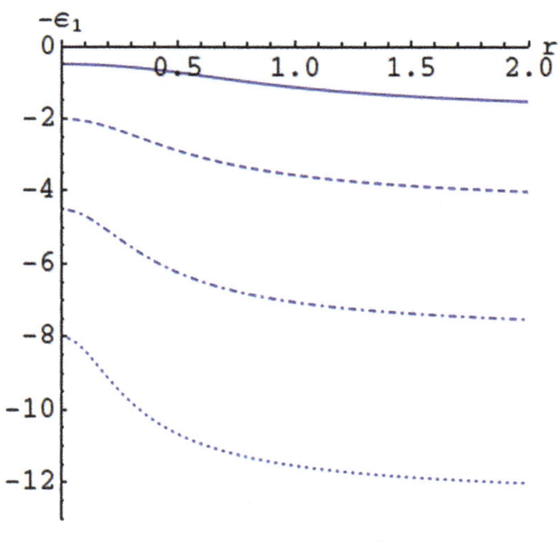

Figure 10.2. Classical energy terms: the scaled energy $-\varepsilon_1 = (L^2/m)E$ versus the scaled internuclear distance $r = (m/L^2)R$ for $Z = 2$ (solid curve), $Z = 3$ (dashed curve), $Z = 4$ (dot-dashed curve) and $Z = 5$ (dotted curve). Reprinted from [9] with permission from the Canadian Journal of Physics.

'Let us summarize the differences from the case of $Z = 1$. The equilibrium range in the case of $Z > 1$ is not half-infinite (as it was for $Z = 1$), but rather it has a lower limit. In terms of the scaled internuclear coordinate $w = z/R$, it is given by equation (6), which yields $-\infty < w < 0$ for $Z = 1$. Thus, the equilibrium range is drastically reduced in the case of $Z > 1$: for $Z = 2$, it is $-1 < w < 0$, for $Z = 3$ it is $-1/2 < w < 0$, and so on.

Each curve in figure 10.2 represents two coinciding energy terms: there is a double degeneracy with respect to the sign of the projection of the muon angular momentum on the internuclear axis. This was also the case for $Z = 1$. However, in distinction to the case of $Z = 1$, now the energy terms start not at some non-zero R_{min}, but at $R_{min} = 0$. This is because the muon would be bound even if the nucleus and electron are at the same point, since their total charge is positive. Another distinction from the case of $Z = 1$ is that the energy of the muon at $R = 0$ is not zero, but rather it is $-(Z-1)^2m/(2L^2)$. As R increases, the energy diminishes approaching $-Z^2m/(2L^2)$ at $R \to \infty$. The binding energies for $Z > 1$ are much greater than for $Z = 1$ at any finite R—for the same reason as stated above: for $Z = 1$ the total charge of the nucleus–electron subsystem is zero, so that the muon is only relatively weakly bound—compared to the case of $Z > 1$, where the total charge of the nucleus–electron subsystem is positive.'

It is important to note that the authors of paper [9] also analyzed the stability of the motion of the muon by considering small oscillations of the radius and the position of the muon orbital plane. They have proven that the motion is stable. These results are represented in appendix A of this book.

The muon revolution frequency behaves analogously to the situation of $Z = 1$. However, for $Z > 1$, its maximum value is Z^2m/L^3. Physically, the latter is the Kepler frequency for the muonic hydrogenic ion, whose nucleus has the charge Z.

The authors of paper [9] derived the following expressions for the revolution frequencies of the muon (Ω) and of the electron (ω):

$$\Omega = \frac{m}{L^3}\nu, \ \nu = \frac{(Z^{2/3}\gamma^4 - 1)^3}{(\gamma^3 - 1)(\gamma^3 + 1)^3} \tag{10.10}$$

$$\omega = \frac{m^{3/2}}{L^3}\frac{\ell_e}{r^{3/2}}. \tag{10.11}$$

In equation (10.11), the equilibrium value of the scaled angular momentum of the electron is:

$$\ell_e = \sqrt{Z}\sqrt{1 - \frac{(1-\gamma)^2\sqrt{1+\gamma+\gamma^2}}{(1-\gamma+\gamma^2)^{3/2}}}. \tag{10.12}$$

By conjoining equations (10.10)–(10.12), the authors of paper [9] obtained the following parametric dependence of the ratio Ω/ω on the ratio L/M:

$$\frac{\Omega}{\omega} = \frac{1}{\sqrt{m}}\frac{\nu r^{3/2}}{\ell_e} \tag{10.13}$$

$$\frac{L}{M} = \frac{\sqrt{m}}{\ell_e\sqrt{r}}. \quad (10.14)$$

Figure 10.3 presents the plot of the ratio Ω/ω versus the ratio L/M for several values of the nuclear charge Z. One can see, that, for instance, for $Z = 5$, the validity condition of the separating the rapid and slow subsystems $\Omega/\omega \gg 1$ requires the ratio L/M to be markedly greater than 34.

The muon radiates a spectral line at the frequency of its revolution $\Omega(R)$. Figure 10.4 shows the dependence of the muonic scaled frequency ν (defined in equation (10.10)) on the scaled separation r of the electron from the nucleus for several values of the nuclear charge Z.

It is seen that there is a red shift counted from the spectral line of the corresponding muonic hydrogenic ion. The authors of paper [9] defined the relative shift as follows:

$$\delta = \frac{\Omega_{\max} - \Omega}{\Omega} \quad (10.15)$$

or equivalently

$$\delta = \frac{1}{\nu(\gamma)} - 1. \quad (10.16)$$

The combination of equations (10.14) and (10.16) represents the parametric dependence of the relative red shift δ on the angular momenta ratio L/M. This dependence is shown in figure 10.5 for several values of the nuclear charge Z.

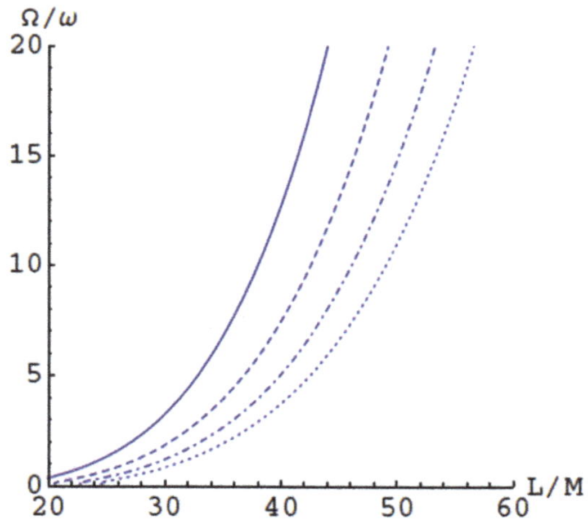

Figure 10.3. The ratio of the muon and electron revolution frequencies Ω/ω versus the ratio of the muon and electron angular momenta L/M for $Z = 2$ (solid curve), $Z = 3$ (dashed curve), $Z = 4$ (dot-dashed curve) and $Z = 5$ (dotted curve). Reprinted from [9] with permission from the Canadian Journal of Physics.

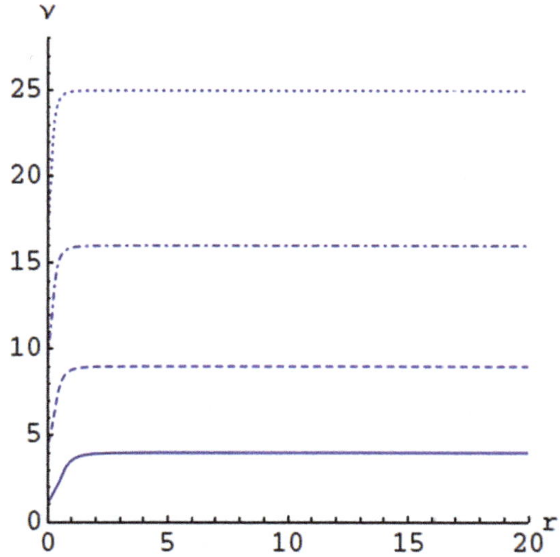

Figure 10.4. The scaled muon revolution frequency $\nu = (L^3/m)\Omega$ versus the scaled internuclear distance $r = (m/L^2)R$, for $Z = 2$ (solid curve), $Z = 3$ (dashed curve), $Z = 4$ (dot-dashed curve) and $Z = 5$ (dotted curve). Reprinted from [9] with permission from the Canadian Journal of Physics.

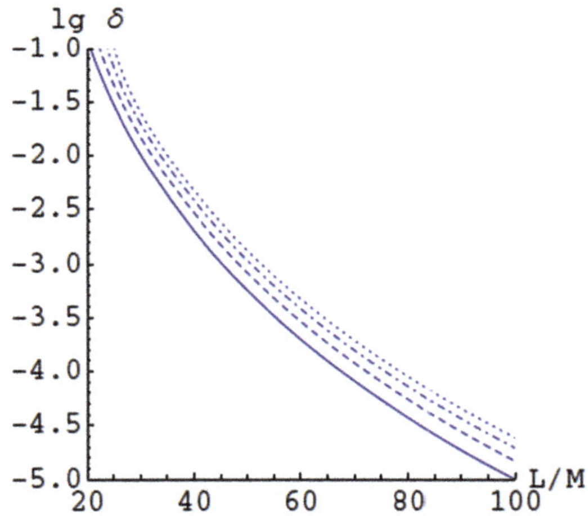

Figure 10.5. Dependence of the relative red shift δ of the spectral lines of the quasi-molecule μZe on the ratio of the muon and electron angular momenta for $Z = 2$ (solid curve), $Z = 3$ (dashed curve), $Z = 4$ (dot-dashed curve) and $Z = 5$ (dotted curve). Reprinted from [9] with permission from the Canadian Journal of Physics.

By employing equations (10.13) and (10.16), the authors of paper [9] plotted also the shift δ versus the muon to electron frequency ratio (figure 10.6) for several values of the nuclear charge Z.

One can see that, as Z grows, the relative red shift diminishes. However, at minimum for the values of Z up to $Z = 5$, the red shift continues to be within the spectral resolution of typical spectrometers.

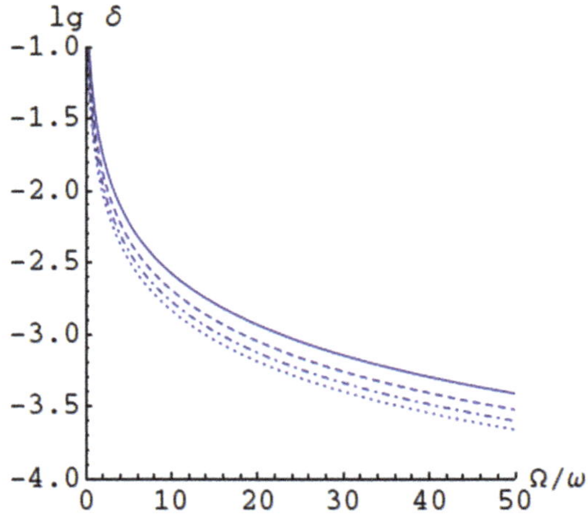

Figure 10.6. Dependence of the relative red shift δ of the spectral lines of the quasi-molecule μ*Ze* on the ratio of the muon and electron frequencies for $Z = 2$ (solid curve), $Z = 3$ (dashed curve), $Z = 4$ (dot-dashed curve) and $Z = 5$ (dotted curve). Reprinted from [9] with permission from the Canadian Journal of Physics.

Table 10.1. Selected parameters of the system for $Z = 2, 3, 4, 5$ and for arbitrary Z. Reprinted from [9] with permission from the Canadian Journal of Physics.

	$Z = 2$	$Z = 3$	$Z = 4$	$Z = 5$	Arbitrary Z
Muon equilibrium range	$-1 < w < 0$	$-1/2 < w < 0$	$-1/3 < w < 0$	$-1/4 < w < 0$	$-1/(Z-1) < w < 0$
Muon maximum energy (at $R = 0$)	$-1/2\ m\ L^{-2}$	$-2\ m\ L^{-2}$	$-9/2\ m\ L^{-2}$	$-8\ m\ L^{-2}$	$-(Z-1)^2/2\ m\ L^{-2}$
Muon minimum energy (at $R \to \infty$)	$-2\ m\ L^{-2}$	$-9/2\ m\ L^{-2}$	$-8\ m\ L^{-2}$	$-25/2\ m\ L^{-2}$	$-Z^2/2\ m\ L^{-2}$
Maximum value of muon frequency	$4\ m\ L^{-3}$	$9\ m\ L^{-3}$	$16\ m\ L^{-3}$	$25\ m\ L^{-3}$	$Z^2 m/L^3$
Critical value of L/M	25	28	30	30	Non-analytical

The authors of paper [9] compiled the following table, recapitulating their findings (table 10.1).

In summary, the authors of paper [9] analyzed quasimolecules μ*Ze* of $Z > 1$ and found that for sufficiently high ratio of the angular momenta of the muon and the electron, the motion of the muon can be treated as the rapid subsystem, and the motion of the electron as the slow subsystem. They showed that the muon binding energy increases as the nuclear charge Z grows.

The authors of paper [9] calculated the red shift of the spectral line, radiated by the muon in the μ*Ze* system, compared to the frequency of the corresponding spectral line of the muonic hydrogenic ion. They demonstrated that as the nuclear charge Z grows, the red shift decreases, but continues to be within the spectral resolution of typical spectrometers. Therefore, the formation of the μZe system can be detected by this red shift.

The authors of paper [9] also emphasized the following.

'We also note that it is possible to go beyond the circular states. From the general analytical results of paper [10], it follows that there are *stable helical states*. For the quasimolecule μZe this means that the muon trajectory would be a helix on the surface of a cone, with the axis coinciding with the line connecting the proton and the electron. In this *helical* state, the muon, while rapidly spiraling on the surface of the cone, oscillates between two end-circles which result from cutting the cone by two parallel planes (very close to each other) perpendicular to its axis. Thus, from the circular state, under a fluctuation the quasimolecule would switch to such stable helical state.'

Finally, it should be underscored that in papers [1, 9] the muon was assumed to have a relatively large angular momentum. This is not out of the realm of experiments. Indeed, muonic hydrogen atoms (μp) are often created in experiments entailing highly-excited states of the muon—see, e.g., works [11, 12]. Besides, in paper [13] it was demonstrated that as the energy of the muon (colliding with the electronic hydrogen atom) increases, the distribution of the principal quantum number of the muon in muonic hydrogen atoms shifts to greater and greater values. This validated the employment of the classical treatment in paper [1] for $Z = 1$, and the case of $Z > 1$ should be similar in this regard.

References

[1] Kryukov N and Oks E 2013 *Can. J. Phys.* **91** 715

[2] Ponomarev L I 1990 *Contemp. Phys.* **31** 219

[3] Nagamine K 2001 *Hyperfine Interact.* **138** 5

[4] Nagamine K and Ponomarev L I 2003 *Nucl. Phys.* A **721** C863

[5] Chelkowsky C, Bandrauk A D and Corkum P B 2004 *Laser Phys.* **14** 473

[6] Guffin J, Nixon G, Javorsek D II, Colafrancesco S and Fischbach E 2002 *Phys. Rev.* D **66** 123508

[7] Oks E 2020 *Res. Astron. Astrophys.* **20** 109

[8] Oks E 2020 *Atoms* **8** 33

[9] Kryukov N and Oks E 2014 *Can. J. Phys.* **92** 1405

[10] Oks E 2000 *J. Phys. B: Atom. Mol. Opt. Phys.* **33** 3319

[11] Measday D F 2001 *Phys. Rep.* **354** 243

[12] Sakimoto K 2010 *Phys. Rev.* A **81** 012511

[13] Garcia J D, Kwong N H and Cohen J S 1987 *Phys. Rev.* A **35** 4068

IOP Publishing

Advances in Physics of Rydberg Atoms and Molecules

Eugene Oks

Chapter 11

One-electron Rydberg quasimolecules in a circularly-polarized high-frequency laser field

In plasmas containing ions of different nuclear charges Z and Z', often the charges Z and Z' come in close proximity to each other. As they share one electron, the resulting physical system is a quasimolecule. When this electron is at the highly-excited energy level, the system represents a one-electron Rydberg quasimolecule.

There have been many classical studies of one-electron Rydberg quasimolecules—see. e.g., papers [1–10]. In particular, they covered the effects of a static magnetic field, or a static electric field, or a low-frequency laser field.

In paper [11], whose results were later represented in chapter 8 of book [12], the authors studied the effect of a high-frequency laser field on one-electron Rydberg quasimolecules. The laser field was assumed to be *linearly-polarized.*

The authors of paper [11] obtained analytical results by employing the generalized method of effective potentials [13–15]. They demonstrated how the configuration of classical energy terms becomes progressively more complex with the growth of the laser amplitude. They also derived the laser-field-caused shift of the frequency, at which the one-electron Rydberg quasimolecule radiates.

Analytical results on one-electron Rydberg quasimolecules in a *circularly-polarized* high-frequency laser field were presented later in paper [16]. Below we overview these additional analytical results.

The circularly-polarized laser field can be presented in the form:

$$F = F(\mathbf{e}_x \cos \omega t + \mathbf{e}_y \sin \omega t). \tag{11.1}$$

In equation (11.1), the unit vectors along the x- and y-axes are denoted as $\mathbf{e}_x$ and $\mathbf{e}_y$. The Hamiltonian can be written as

$$H = H_0 + F\rho \cos(\varphi - \omega t) = H_0 + F\rho(\cos \varphi \cos \omega t + \sin \varphi \sin \omega t) \tag{11.2}$$

doi:10.1088/978-0-7503-3939-1ch11

in distinction to the case of the linearly-polarized laser field, where the Hamiltonian was

$$H = H_0 + zF\cos\omega t,\; H_0 = \frac{1}{2}\left(p_z^2 + p_\rho^2 + \frac{p_\varphi^2}{\rho^2}\right) - \frac{Z}{r} - \frac{Z'}{r'}. \tag{11.3}$$

We note that both in equation (11.2) and in equation (11.3), H_0 is the same and the cylindrical coordinates (ρ, φ, z) with the z-axis perpendicular to the laser polarization plane have been used.

The authors of paper [16] introduced the following notations:

$$V = F\rho\cos\varphi,\quad W = F\rho\sin\varphi. \tag{11.4}$$

Then they employed the generalized formalism of effective potentials [12, 13] for the circular-polarization case:

$$U_0 = \frac{1}{4\omega^2}([V, [V, H_0]] + [W, [W, H_0]]) = \frac{F^2}{4\omega^2} \tag{11.5}$$

$$U_1 = \frac{1}{4\omega^4}([[V, H_0], [[V, H_0], H_0]] + [[W, H_0], [[W, H_0], H_0]]) + \frac{-1}{2\omega^3} [[V, H_0], [W, H_0]]. \tag{11.6}$$

The 0th order term given by equation (11.5) is irrelevant to the dynamics of the system because it does not depend on coordinates. The first order term given by equation (11.6) yields:

$$U_1 = \frac{F^2}{4\omega^4}\left(\frac{3L^2}{\rho^4} + Z\frac{z^2 - 2\rho^2}{(z^2 + \rho^2)^{5/2}} + Z'\frac{(R - z)^2 - 2\rho^2}{((R - z)^2 + \rho^2)^{5/2}}\right). \tag{11.7}$$

After using the scaled quantities

$$w = \frac{z}{R},\; v = \frac{\rho}{R},\; \varepsilon = -\frac{R}{Z}E,\; b = \frac{Z'}{Z},\; \ell = \frac{L}{\sqrt{ZR}},\; r = \frac{Z}{L^2}R,\; \theta = \frac{F}{\omega^2 R}. \tag{11.8}$$

The authors of paper [16] derive the following scaled electron energy

$$\begin{aligned}\varepsilon = &\frac{1}{\sqrt{w^2 + v^2}} + \frac{b}{\sqrt{(1 - w)^2 + v^2}} - \frac{\ell^2}{2v^2} - \frac{w^2 - 2v^2}{(w^2 + v^2)^{5/2}}\frac{\theta^2}{4} \\ &- b\frac{(1 - w)^2 - 2v^2}{((1 - w)^2 + v^2)^{5/2}}\frac{\theta^2}{4} - \frac{3\ell^2}{v^4}\frac{\theta^2}{4}.\end{aligned} \tag{11.9}$$

By equating to zero the partial derivatives of the scaled electron energy with respect to w and v, they obtained the equilibrium conditions. From one of these conditions they found the equilibrium value of the scaled angular momentum:

$$\ell = \frac{v^3}{\sqrt{v^2 + 3\theta^2}} \sqrt{\frac{(w^2 + v^2)^2 - \frac{3}{4}(3w^2 - 2v^2)\theta^2}{(w^2 + v^2)^{7/2}} + b\frac{((1 - w)^2 + v^2)^2 - \frac{3}{4}(3(1 - w)^2 - 2v^2)\theta^2}{((1 - w)^2 + v^2)^{7/2}}}. \tag{11.10}$$

The other equilibrium condition they represented in the form:

$$w\frac{(w^2 + v^2)^2 - 3\left(\frac{w^2}{4} - v^2\right)\theta^2}{(w^2 + v^2)^{7/2}} = b(1 - w)\frac{((1 - w)^2 + v^2)^2 - 3\left(\frac{(1 - w)^2}{4} - v^2\right)\theta^2}{((1 - w)^2 + v^2)^{7/2}}. \tag{11.11}$$

Figure 11.1 presents the equilibrium curve in the plane of the coordinates w and v for the scaled laser amplitude $\theta = 0.2$. The shaded parts correspond to the forbidden region, where formally calculated square of the angular momentum $\ell^2 < 0$. Therefore, the equilibrium curves within the forbidden region should be disregarded.

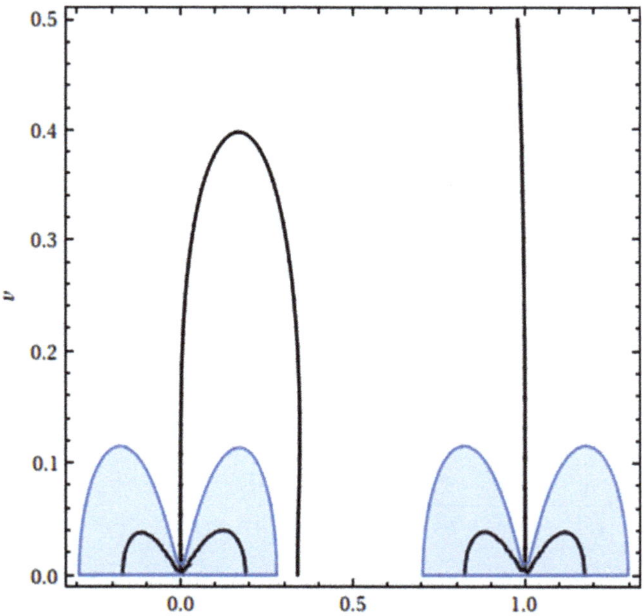

Figure 11.1. Equilibrium plot in the (w, v)-plane for $b = 3$ and $\theta = 0.2$. The forbidden region, corresponding to the points where $\ell^2 < 0$, is shaded. [16] (2020). With permission of Springer.

The authors of paper [16] wrote (we updated the numeration of the equations):

'With the master equations (11.9), (11.10) and (11.11), we proceed as in the linear case: defining $r = 1/\ell^2$ and $\varepsilon_1 = \varepsilon/r$ and numerically solving (11.11) for v and then substituting it into r and ε_1, we obtain the parametric dependence $\varepsilon_1(r)$ representing the scaled energy terms, with the parameter w running over the allowed range determined by $0 < w < w_1$, $b/(b + 1) < w < 1$ (for $\theta < \theta_c$); for $\theta_c < \theta < \theta_{10}$, w_1 is changed into the point with the maximum w on the left branch, determined numerically, and for $\theta > \theta_{10}$, it is changed into the point of intersection of the branch with the forbidden region, except when its w-coordinate is less than the maximum w for the branch.'

Figure 11.2 depicts the scaled energy terms for the scaled laser amplitude of the laser field $\theta = 0.3$ in comparison to the terms at the absence of the laser field (i.e., for $\theta = 0$). One can see that for $\theta = 0.3$, the scaled energy terms significantly differ from the unperturbed ones, but no new terms appear—in distinction to the case of the linearly-polarized laser field presented in [11, 12].

Finally the authors of paper [16] studied the red shift of the corresponding spectral lines. They denoted:

$$\Omega = \sqrt{\frac{Z}{R^3}}\tilde{\Omega}, \quad \tilde{\Omega} = \frac{\ell}{v^2} \tag{11.12}$$

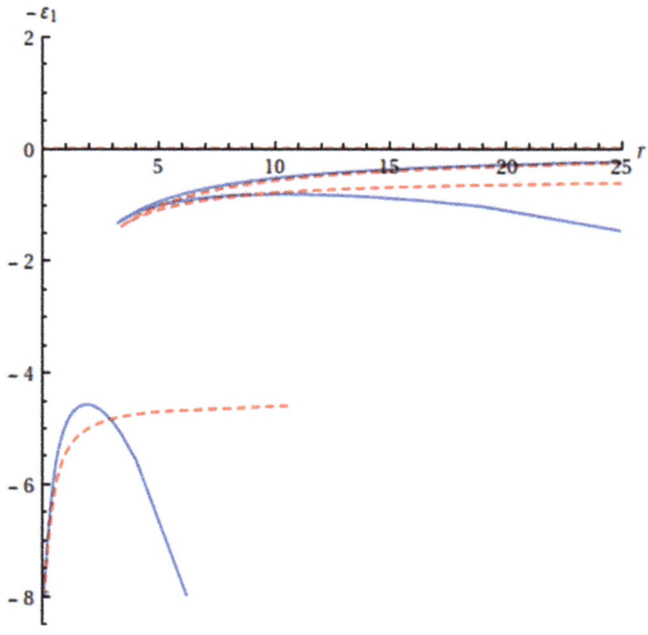

Figure 11.2. The plot of the scaled energy terms $-\varepsilon_1(r)$ (with r on the horizontal axis and ε_1 on the vertical) for the scaled amplitude of the laser field $\theta = 0.3$ in the case of circular polarization, with $b = 3$, shown in blue, solid curves, against the terms for $\theta = 0$, with $b = 3$, shown in red, dashed curves. [16] (2020). With permission of Springer.

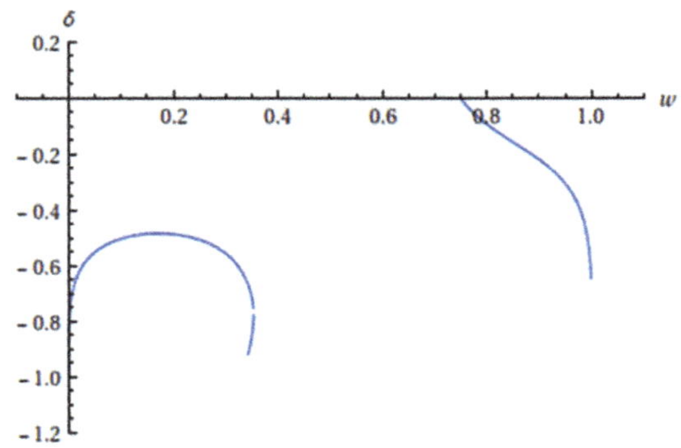

Figure 11.3. The relative red shift of the radiation frequency of the electron versus its scaled axial coordinate for $b = 3$ and $\theta = 0.3$ in the case of circular polarization. [16] (2020). With permission of Springer.

In equation (11.12), the wave (the tilde) above Ω signifies the scaled frequency. The relative frequency shift δ is defined as follows:

$$\delta = \frac{\Omega - \Omega_0}{\Omega_0} = \frac{\Omega}{\Omega_0} - 1 = \frac{\tilde{\Omega}}{\tilde{\Omega}_0} - 1 = \frac{\ell}{\ell_0}\frac{v_0^2}{v^2} - 1. \tag{11.13}$$

In equation (11.13), the quantities with the subscript '0' correspond to the case of the zero laser field ($\theta = 0$); for the quantity v there was taken the equilibrium value found from equation (11.11).

The authors of paper [16] further wrote (we updated the numeration of the equations and figures):

'As in the case of linear polarization, we studied the shift of the radiation frequency caused by a high-frequency circularly-polarized laser field. Following the same procedure to derive the formula for the shift as in the linear-polarization case and using the result in (11.13), we plot the relative shift of the frequency in figure 11.3. It is seen that in the case of circular polarization, red shift is observed ($\delta < 0$)—the effect opposite to that in the case of linear polarization, where blue shift was observed ($\delta > 0$).'

In summary, the authors of paper [16] complemented the results from their paper [11] (where they considered one-electron Rydberg quasimolecules in a *linearly-polarized* high-frequency laser field) by analyzing one-electron Rydberg quasimolecules in a *circularly-polarized* high-frequency laser field). They found the following two major distinctions from the case of the linear polarization.

First, in the case of the linear polarization, as the laser amplitude grows, the number of the energy terms increases. However, in the case of the circular polarization no additional energy terms appear.

Second, in the case of the linear polarization, the electron radiation frequency shifted to the blue. However, in the case of the circular polarization, the electron radiation frequency shifted to the red. When the laser amplitude is known, one can perform the experimental determination of the separation between the nucleus and the electron orbital plane by measuring the frequency shift of the radiation.

References

[1] Oks E 2000 *Phys. Rev. Lett.* **85** 2084

[2] Oks E 2000 *J. Phys. B: Atom. Mol. Opt. Phys.* **33** 3319

[3] Oks E 2001 *Phys. Rev.* E **63** 057401

[4] Flannery M R and Oks E 2006 *Phys. Rev.* A **73** 013405

[5] Kryukov N and Oks E 2011 *Intern. Rev. Atomic Mol. Phys.* **2** 57

[6] Kryukov N and Oks E 2012 *Can. J. Phys.* **90** 647

[7] Kryukov N and Oks E 2012 *Intern. Rev. Atomic Mol. Phys.* **3** 113

[8] Kryukov N and Oks E 2013 *J. Phys. B: Atom. Mol. Opt. Phys.* **46** 245701

[9] Kryukov N and Oks E 2014 *Eur. Phys. J.* D **68** 171

[10] Kryukov N and Oks E 2013 *Intern. Rev. Atomic and Mol. Phys.* **4** 121

[11] Kryukov N and Oks E 2019 *Intern. Rev. Atomic Mol. Phys.* **9** 1

[12] Oks E 2020 *Analytical Advances in Quantum and Celestial Mechanics: Separating Rapid and Slow Subsystems* (Bristol: IOP Publishing)

[13] Nadezhdin B B *Radiatsionnye i Relativistskie Effekty v Atomakh i Ionakh* (*Radiative and Relativistic Effects in Atoms and Ions*) (Moscow: Scientific Council of the USSR Academy of Sciences on Spectroscopy) p 222 (in Russian)

[14] Kapitza P L 1951 *Sov. Phys. JETP* **21** 588

[15] Kapitza P L 1951 *Uspekhi Fiz. Nauk* **44** 7

[16] Kryukov N and Oks E 2020 *Eur. Phys. J.* D **74** 82

Chapter 12

Muonic-electronic helium and helium-like ions

12.1 Brief overview of the classical analytical solution

Muonic atoms and molecules, that is, microscopic systems, where a muon substituted one electron, have been studied both for fundamental and practical reasons. In particular, the applications include muon-catalyzed fusion (see, e.g. [1–3] and references therein), the possibility of controlling nuclear processes by a laser [4], the hunt for strongly interacting massive particles (SIMPs) that were suggested in the past as possible candidates for dark matter (see, e.g. [5] and references therein), though by now this hypothesis is practically dismissed.

In book [6] (chapter 4) and paper [7] we analyzed muon-electronic heliumic systems (that is, helium atoms or helium-like ions) where both the muon and the electron are in highly-excited (Rydberg) states. In particular, a circular Rydberg state was supposed to be the state of the muon. However, as for the shape of the orbit of the electron and the orientation of the electronic orbit with respect to the muon orbital plane, no supposition was made.

In this section we very briefly reiterate only those main results from book [6] (chapter 4) and paper [7] that are necessary for understanding the two other sections of this chapter.

The primary point is that in such muonic-electronic heliumic systems, the electronic motion is characterized by a much lower frequency than the muonic motion. Therefore, it was legitimate to apply the analytical technique employing the separation of rapid and slow subsystems [6, 8, 9].

One of the main results was the expression for the ratio of the muon frequency Ω to the Kepler frequency of the electron ω_K—the ratio that has to be much greater than unity for the validity of separating the rapid and slow subsystems:

$$\Omega/\omega_K = m_{\mu r} Z^2 n_e^3/\left[m_e (Z - 1)^2 n_\mu^3\right] >> 1 \qquad (12.1)$$

doi:10.1088/978-0-7503-3939-1ch12

if

$$n_\mu / n_e << (m_{\mu r}Z^2)^{1/3} / [m_e(Zm - 1)^2]^{1/3}. \qquad (12.2)$$

In equations (12.1) and (12.2), Z is the nuclear charge; $m_{\mu r}$ and m_e are the muon and electron masses, respectively; n_μ and n_e are the principal quantum numbers of the muon and electron states, respectively.

For He atoms, ($Z = 2$), equation (12.2) leads to

$$n_\mu / n_e < 9, \qquad (12.3)$$

so that equation (12.1) simplifies to:

$$\Omega / \omega_K = 804.\ 4\ (n_\mu/n_e)^3. \qquad (12.4)$$

For $Z \gg 1$, equation (12.2) leads to

$$n_\mu / n_e < 6, \qquad (12.5)$$

so that equation (12.1) simplifies to:

$$\Omega / \omega_K = 206.\ 8\ (n_\mu / n_e)^3. \qquad (12.6)$$

The author of works [6, 7] wrote:

'Thus, it is seen that, e.g., at $n_\mu/n_e \sim 1$, the frequency of the revolution of the muon exceeds the Kepler frequency of the electron by two or three orders of magnitude (obviously, the ratio Ω/ω_K is even greater for $n_\mu/n_e < 1$). So, indeed the pair 'nucleus-muon' constitute a rapid subsystem, while the electron—a slow subsystem.'

For determining the motion of the electron, it is necessary to average over the rapid susbsystem. This leads to the following physical situation.

From the point of view of the electron, instead of the muon and the nucleus there are two concentric rings of the radii R_μ (for the muon) and R_{nucl} (for the nucleus). The muon charge is uniformly distributed over the muonic ring and the nuclear charge is uniformly distributed over the nucleonic ring.

After some calculations, in works [6, 7] it was shown that the effective potential energy for the electron is as follows:

$$U_{eff} = -(Z - 1)\,e^2 / r - e^2 R_\mu^2\,(3\cos^2\theta - 1) / (4r^3). \qquad (12.7)$$

This effective potential has a celestial analogy. For a satellite moving around the oblate Earth, the corresponding potential is mathematically equivalent to the one from equation (12.7). The solution of the latter problem presented, for example, in book [10] demonstrates that the satellite elliptical orbit of the satellite is engaged simultaneously in two types of the precession, but the shape of the orbit does not vary in time.

So, in the muonic-electronic heliumic system, the electron elliptical orbit is engaged simultaneously in two types of the precession, but the shape of the orbit does not vary in time.

The first one is the precession of the orbit in its plane. This precession occurs with the angular frequency

$$\omega_{\text{precession in plane}} = (3/8)[(Z - 1)/Z^2](1 - 5\cos^2 i)\,[n_\mu/(l_e + 1/2)]^4(m_{er}/m_{\mu r})^2\omega_K. \quad (12.8)$$

In equation (12.8), ω_K is the Kepler frequency of the highly-excited electron, l_e is the angular momentum quantum number of the electron, and i is the inclination, which is the angle between the planes of the muonic and electronic orbits. In particular, for $i = 63.4$ degrees, one obtains

$1–5\cos^2 i = 0$, so that the corresponding frequency is zero: no precession within the orbital plane.

As for the second precession, it is the precession of the electron orbital plane about the axis of the symmetry of the muonic and nucleonic 'rings' with the following frequency:

$$\omega_{\text{precession in plane}} = (3/4)[(Z - 1)/Z^2](\cos^2 i)[n_\mu/(l_e + 1/2)]^4(m_{er}/m_{\mu r})^2\omega_K. \quad (12.9)$$

For $i = 90$ degrees, this precession frequency vanishes. This is the configuration where the orbital planes of the muon and electron are perpendicular to each other. In this configuration, the electron orbital plane does not precess about the axis of the symmetry of the muonic and nucleonic 'rings'.

12.2 Counterintuitive possibility of the linear Stark effect

While hydrogenic atoms/ions exhibit the linear Stark effect, helium atoms and helium-like ions (hereafter, heliumic systems) do not demonstrate the linear Stark effect in a relatively weak external electric field—according to the existing paradigm (see, e.g., the textbooks [11, 12]). The overwhelming majority of states of hydrogenic systems have a non-zero value of the averaged electric dipole moment, which is the reason for the linear Stark effect in these systems, as is well-known. We recall that the averaging is performed in the quantum formalism over unperturbed wave functions or in the classical formalism (relevant to highly-excited states) over the unperturbed electronic orbit.

In distinction, according to the existing paradigm, for heliumic systems the averaged electric dipole moment is zero. Therefore, no linear Stark effect in a weak field should be exhibited by heliumic systems—even if one of the electrons is highly-excited. In the classical description, the unperturbed elliptical orbit of the highly-excited electron precesses. The precession is caused by the fact that at relatively small distances from the nucleus of charge Z, the effective potential deviates from the Coulomb potential of the effective nuclear charge $(Z–1)$. Of course, the unperturbed elliptical orbit of the electron would have a non-zero electric dipole moment; however, after averaging over the precession it becomes zero.

In paper [13], which we overview in this section, the author reconsidered the same physical system as in paper [7], but from a different point of view. Namely, he showed that the averaged electric dipole moment differs from zero in two states of the system. Therefore, the system in these states would respond even to a weak electric field by the linear Stark effect.

Besides, the author of paper [13] showed that by using the external electric field one can control the electronic orbit precession. He pointed out in particular that the precession can be stopped by applying a certain value of the external electric field and calculated this critical value. Below are some details.

The author of paper [13] employed as the starting point the setup from paper [7] where the radius R_μ of the muonic circular orbit and the major semi-axis R_e of the electronic elliptical orbit satisfy the inequality

$$R_e \gg R_\mu. \tag{12.10}$$

The orbital quantities R_μ and R_e have the following principal quantum numbers as their quasiclassical equivalents

$$n_e = [(Z - 1)R_e/a_0]^{1/2}, \; n_\mu = [ZR_\mu m_{\mu r}/(a_0 m_{er})]^{1/2}. \tag{12.11}$$

In equation (12.11) a_0 is the Bohr radius, while

$$m_{\mu r} = m_\mu(m_e + M_{nucl})/(m_e + m_\mu + M_{nucl}) \tag{12.12}$$

and

$$m_{er} = m_e (m_\mu + M_{nucl}) / (m_e + m_\mu + M_{nucl}) \tag{12.13}$$

are the muon and of the electron reduces masses, respectively. We recall that $m_\mu/m_e = 206.8$.

The author of paper [13] showed that the separation of the rapid and slow subsystems is justified if

$$R_\mu/R_e < (m_e/m_\mu)^{1/3} [Z/(Z - 1)]^{1/3}. \tag{12.14}$$

Quasiclassically, the condition (12.14) is equivalent to

$$n_\mu/n_e < (m_{\mu r}Z^2)^{1/3} / [m_{er}(Z - 1)^2]^{1/3}. \tag{12.15}$$

So, the electronic orbit get engaged in the two precessions explained in section 12.1 above. Quasiclassical expressions for the frequencies of these two precessions were given above by equations (12.8) and (12.9). The classical counterparts of equations (12.8) and (12.9) are as follows

$$\omega_{\text{precession in plane}} = \left[3(Z - 1)m_{er}{}^2 e^4 R_\mu{}^2/(8M_e{}^4)\right](1 - 5\cos^2 i), \tag{12.16}$$

$$\omega_{\text{precession of plane}} = -\left[3(Z - 1)m_{er}{}^2 e^4 R_\mu{}^2/(4M_e{}^4)\right]\cos i. \tag{12.17}$$

Figure 12.1 presents the dependence of the ratio of these two frequencies on the inclination i of the electron orbital plane. One can see that, except for the vicinity of $|i| = \arccos(1/5^{1/2}) = 1.107$ rad $= 63.43°$, these two frequencies (by the absolute value) are of the same order of magnitude. The author of paper [13] called the angle arcos $(1/5^{1/2})$, which is the solution of the equation

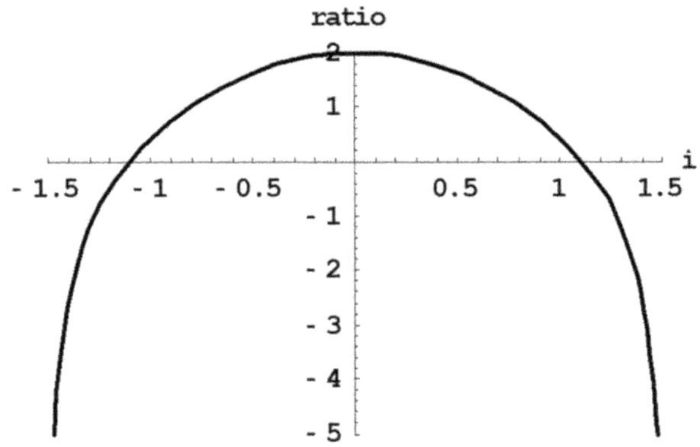

Figure 12.1. The ratio of the frequency of the precession in plane (from equation (12.16)) to the frequency of the precession of the plane (from equation (12.17)) versus the inclination i of the orbital plane of the electron. Reprinted from [13] with permission from MDPI.

$$1 - 5\cos^2 i = 0, \tag{12.18}$$

the critical angle and denoted it as i_c.

The primary result of paper [13] is the following. The frequency of the precession of the electron orbit in the orbital plane becomes zero as the inclination angle has either one of the following two values

$$i = \pm i_c \approx \pm 1.107\ \text{rad} = \pm 63.43°. \tag{12.19}$$

Due to the absence of the precession within the orbital plane at these inclination angles, *the dipole moment does not average to zero.*

At $i = \pm i_c$, the frequency of the precession of the plane of is the following, according to equation (12.17):

$$\omega_{\text{precession of plane}} = -\left[3(Z - 1)m_{er}{}^2 e^4 R_\mu{}^2/(5^{1/2}4M_e{}^4)\right]. \tag{12.20}$$

The author of paper [13] wrote (we updated the numeration of the equations):

'This precession does not eliminate the dipole moment: in the course of this precession, the dipole moment precesses with the frequency $\omega_{\text{precession of plane}}$. Therefore, the projection of the dipole moment on the axis of the precession remains constant. Thus, at the two values of the inclination given by equation (12.19), the system should exhibit the *linear* Stark effect *even in a vanishingly small electric field.*

We emphasize that a sufficiently high electric field would intermix energy levels of a heliumic system and thus cause the linear Stark effect. However, the new result presented above is the linear Stark effect in a heliumic system even in an arbitrarily small electric field. This is a *counter-intuitive result.*'

The projection of the electric dipole moment on the axis of symmetry of the system (which is the axis of the precession, chosen as the z-axis) is the following (according to appendix B of this book)

$$d_z = e < z_e > \left[M_n + m_\mu + (Z - 1)m_e\right]/(M_n + m_\mu + m_e). \tag{12.21}$$

In equation (12.21), $<z_e>$ is the average value of the electronic z-coordinate:

$$<z_e> = <r_e> \cos\theta = <r_e> \sin i = \pm 2 < r_e > /5^{1/2}. \quad (12.22)$$

In equation (12.22), $<r_e>$ is the absolute value of the average radius-vector of the electron. The corresponding expression for $<r_e>$ is (see, e.g. [14–17]):

$$<r_e> = [3e^2(Z-1)/(4|E_e|)]\{1 - 2|E_e|M_e^2/[(Z-1)^2 m_{er}e^4]\}^{1/2}. \quad (12.23)$$

In equation (12.23), E_e is the electron unperturbed energy. The quasiclassical equivalent of equation (12.23) is:

$$<r_e> = 3\hbar^2 n_e[n_e^2 - (l_e + 1/2)^2]^{1/2} / [2(Z-1)m_{er}e^2]. \quad (12.24)$$

By combining equations (12.21)–(12.23), the author of paper [13] got:

$$d_z = \pm\{[M_n + m_\mu + (Z-1)m_e]/(M + m_\mu + m_e)\}$$
$$[3e^3(Z-1)/(5^{1/2}\, 2|E_e|)]\{1 - 2|E_e|M_e^2/[(Z-1)^2 m_{er}e^4]\}^{1/2}. \quad (12.25)$$

The quasiclassical equivalent of equation (12.25) is:

$$d_z = \pm\{[M_n + m_\mu + (Z-1)m_e]/(M + m_\mu + m_e)\}3\hbar^2 n_e$$
$$[n_e^2 - (l_e + 1/2)^2]^{1/2}/[5^{1/2}(Z-1)m_{er}e^2]. \quad (12.26)$$

When a uniform electric field **F** is applied, being directed along the z-axis, the electron orbital plane undergoes yet another precession. The author of paper [13] obtained the following formula for the frequency ω_F of this precession by employing the corresponding results from book [14]:

$$\omega_F = \pm(3eF/2)\{[M_n + m_\mu + (Z-1)m_e]/[(M + m_\mu + m_e)]5^{1/2}m_{er}|E_e|\}^{1/2}. \quad (12.27)$$

The quasiclassical counterpart of equation (12.27) is

$$\omega_F = \pm[1/m_e + (Z-1)/(M + m_\mu)]3n_e[n_e^2 - (l_e + 1/2)^2]^{1/2}$$
$$F/[5^{1/2}(Z-1)e]. \quad (12.28)$$

The two possible directions of the precession are manifested by the two possible signs in equation (12.28). For example, the plus sign in equations (12.27) and (12.28) corresponds to the inclination angle equal to its critical value $i_c \approx +1.107$ rad = +63.43°.

The combined frequency of the orbital plane precession is the following:

$$[1/m_e + (Z-1)/(M + m_\mu)]4Z^2 m_{\mu r}^2(l_e + 1/2)^4 n_e^4[n_e^2 - (l_e + 1/2)^2]^{1/2}\hbar^4$$
$$F/[(Z-1)^4 n_\mu^4 m_{er}^3 e^5]. \quad (12.29)$$

From equation (12.29) it is seen that the orbital plane precession can be *electrically-controlled.*

The author of paper [13] wrote:

'In particular, for the critical value of the angle of the inclination $i_c \approx +1.107$ rad = +63.43°, there exists a critical value of the electric field F_{crit}, such that there would be no precession: $\omega = 0$. In other words, the electric field F_{crit} would 'kill' the precession and make the elliptical orbit stationary. This is another *counter-intuitive result.*'

For determining F_{crit}, the author of paper [13] first obtained the ratio $|\omega_F|/\omega_{precession\ of\ plane}$, as follows:

$$|\omega_F|/\omega_{precession\ of\ plane} = [1/m_e + (Z-1)/(M+m_\mu)]4Z^2m_{\mu r}^2(l_e+1/2)^4n_e^4[n_e^2-(l_e+1/2)^2]^{1/2} \quad (12.30)$$
$$h^4F/[(Z-1)^4n_\mu^4m_{er}^3e^5].$$

Since $1/m_e >> (Z—1)/(M+m_\mu)$, equation (12.30 $0 < E_s \leqslant 1$) can be simplified (after substituting $m_{\mu r} \approx m_\mu$ and $m_{er} \approx m_e$,) to the following:

$$|\omega_F|/\omega_{precessionofplane} \approx [4Z^2/(Z-1)^4](m_\mu/m_e)^2(n_e/n_\mu)^4(l_e+1/2)^4 [n_e^2-(l_e+1/2)^2]^{1/2}F/F_{at}. \quad (12.31)$$

In equation (12.31)

$$F_{at} = m_e^2e^5/\hbar^4 \quad (12.32)$$

is the atomic unit of the electric field.

Then the author of paper [13] equated to unity the ratio $|\omega_F|/\omega_{precession\ of\ plane}$ from equation (12.31) and got

$$F_{crit}/F_{at} \approx (Z-1)^4(m_e/m_\mu)^2(n_\mu/n_e)^4/\left\{4Z^2(l_e+1/2)^4[n_e^2-(l_e+1/2)^2]^{1/2}\right\}. \quad (12.33)$$

By combining equation (12.33) with equation (12.2) for the validity condition, he found the following:

$$F_{crit}/F_{at} \ll Z^{2/3}(Z-1)^{1/3}(m_e/m_\mu)^{2/3}/\left\{4(l_e+1/2)^4[n_e^2-(l_e+1/2)^2]^{1/2}\right\} \quad (12.34)$$
$$\sim [Z^{2/3}(Z-1)^{1/3}/(4n_e^5)](m_e/m_\mu)^{2/3} \ll 1.$$

One can see that it would be sufficient to apply some electric field that is significantly lower than the atomic unit for stopping the precession of the electron elliptical orbit and thus forcing it to be stationary. For instance, if one considers the satiation where $n_e \sim n_\mu \gg 1$ and $l_e = n_e - k$ (k being the integer, such that $k \ll n_e$), then equation (12.33) would lead to

$$F_{crit}/F_{at} \sim 1.5 \times 10^{-6}/[n_e^{9/2}(2k-1)^{1/2}]. \quad (12.35)$$

For example, for $n_e \sim 10$ and $k = 2$, equation (12.35) yields $F_{crit} \sim 10^{-1}$ V cm^{-1}. For another example where $n_e \gg n_\mu \gg 1$, the author of paper [13] obtained even significantly lower values of the critical field F_{crit}:

$$F_{crit}/F_{at} \ll 1.5 \times 10^{-6}/[n_e^{9/2}(2k-1)^{1/2}]. \quad (12.36)$$

In summary, the main results of paper [13] are the following. First, two values of the inclination angle i were found, such that the precession of the electron orbit within its plane ceases to exist.

Second, it was pointed out that for these values of the inclination angle i, the muonic-electronic heliumic system would respond by the *linear* Stark effect even if the applied electric field **F** was very weak. This result is contrary to the existing paradigm for heliumic systems in a weak electric field. Therefore, this result is absolutely *counterintuitive*.

Third, by calculating the combined precession frequency of the electron orbit in the field **F**, it was shown that the precession can be *electrically-controlled*.

Fourth, the existence of a critical field F_{crit} was demonstrated, such that it would terminate the precession: the electron elliptical orbit would become stationary—the second *counterintuitive* result. The typical value of F_{crit} was found to be of the order or smaller than 1 V cm^{-1}.

At the end, the author of paper [13] emphasized the following:

> Finally we note that for the usual helium atom or helium-like ion, where both light particles are electrons, the linear Stark effect in a vanishingly small electric field could be still possible, provided that the size of the major semi-axis of the elliptical orbit of the outer electron is much greater than the radius of the circular orbit of the inner electron.

12.3 Extension from circular to low-eccentricity orbits of the muon and the nucleus

As only circular orbits of the muon and the nucleus were assumed in paper [7], the open question was whether the results would significantly change by taking into account the ellipticity of their orbits. This question was addressed in paper [18], which we overview in this section.

For the motion of the electron, the authors of paper [18] obtained an extra term in the effective potential—the term depending on the eccentricity ε of the muon and nucleus orbits. For the case of the planar geometry, they demonstrated that, as the outcome, the frequency of the precession of the orbit of the electron acquires an extra contribution that depends on ε. They showed that the latter contribution, by the order of magnitude, can become comparable with the precession frequency for the case of the circular orbits of the muon and nucleus. Below are some details. For a Coulomb-like potential

$$U = -\frac{\alpha}{r} \tag{12.37}$$

with $\alpha = Ze^2$, the well-known equation of the motion (in the plane of the orbit) is

$$\frac{p}{r} = 1 + \varepsilon \cos \varphi. \tag{12.38}$$

In equation (12.38)

$$p = \frac{L^2}{m_r \alpha}, \ \varepsilon = \sqrt{1 + \frac{2EL^2}{m_r \alpha^2}}. \tag{12.39}$$

Here E and L are the energy and the angular momentum of the particle, respectively; m_r is the reduced mass of the pair 'Coulomb center—particle'.

The authors of paper [18] considered the situation where the eccentricity of the muon and nucleus orbits is relatively small: $\varepsilon \ll 1$. In this situation one has

$$r = \frac{p}{1 + \varepsilon \cos \varphi} \approx p(1 - \varepsilon \cos \varphi) = r_0(1 - \varepsilon \cos \varphi). \tag{12.40}$$

In equation (12.40), the radius of the circular orbits of the muon and electron at $\varepsilon = 0$ is denoted as r_0.

It should be recalled that for $\varepsilon = 0$, the effective potential for the highly-excited electron has been as follows:

$$U_{\text{eff}}^{(0)} = -\frac{(Z-1)e^2}{r} - \frac{e^2(R_{\mu 0}^2 - ZR_{\text{nucl}0}^2)}{4r^3}(3\cos^2\theta - 1). \tag{12.41}$$

In the situation considered in paper [18] ($\varepsilon \ll 1$), the time dependence of the quantities R_μ and R_{nucl} is the following:

$$R_\mu = R_{\mu 0}[1 - \cos(\Omega t)], \qquad R_{\text{nucl}} = R_{\text{nucl}0}[1 - \cos(\Omega t)]. \tag{12.42}$$

In equation (12.42),

$$\Omega = \sqrt{\frac{Ze^2}{m_{\mu r}(R_{\mu 0} + R_{nucl0})^3}} \approx \sqrt{\frac{Ze^2}{m_{\mu r}R_{\mu 0}^3}} \tag{12.43}$$

is the revolution frequency of the nucleus-muon pair and

$$m_{\mu r} = \frac{m_\mu m_{\text{nucl}}}{m_\mu + m_{\text{nucl}}} \tag{12.44}$$

is the reduced mass of this pair. By combining equations (12.41) and (12.42), the authors of paper [18] got the time-dependent 'potential' as follows:

$$U(t) = U_{\text{eff}}^{(0)} + \frac{e^2(R_{\mu 0}^2 - ZR_{\text{nucl}0}^2)}{2r^3}(3\cos^2\theta - 1)\varepsilon\cos\Omega t \equiv U_{\text{eff}}^{(0)} + W(r, \cos\theta)\cos\Omega t. \tag{12.45}$$

Then they applied the technique of effective potentials to treat the time-dependent term $W(r, \cos\theta)\cos\Omega t$ and used for this purpose the following unperturbed Hamiltonian

$$H_0 = \frac{p^2}{2m_{er}} - \frac{(Z-1)e^2}{r}. \tag{12.46}$$

In equation (12.46),

$$m_{er} = \frac{m_e(m_\mu + m_{nucl})}{m_e + m_\mu + m_{nucl}} \tag{12.47}$$

is the reduced mass of the electron in this system, the reduced mass being practically unity for any Z. It should be also noted that R_{nucl0} can be neglected in the effective potential since $R_{\mu 0}/R_{nucl0} \gg 1$.

The effective potential of the order zero is

$$U_0 = \frac{1}{4\Omega^2}[W, [W, H_0]] = \frac{9\varepsilon^2 e^4 R_{\mu 0}^4 (1 - 2\cos^2\theta + 5\cos^4\theta)}{16 m_{er} \Omega^2 r^8}. \tag{12.48}$$

In equation (12.48), the notation $[P, Q]$ stands for the Poisson brackets; the definition of the quantity W was given in equation (12.45). By combining equations (12.43) and (12.48), the authors of paper [18] obtained

$$U_0 = \frac{9\varepsilon^2 m_{\mu r} e^2 R_{\mu 0}^4 (1 - 2\cos^2\theta + 5\cos^4\theta)}{16 m_{er} Z r^8} \tag{12.49}$$

thus leading to the total effective potential as follows:

$$\begin{aligned} U_{eff} = U_{eff}^{(0)} + U_0 = &-\frac{(Z-1)e^2}{r} - \frac{e^2 R_{\mu 0}^2}{4r^3}(3\cos^2\theta - 1) \\ &+ \frac{9\varepsilon^2 m_{\mu r} e^2 R_{\mu 0}^4 (1 - 2\cos^2\theta + 5\cos^4\theta)}{16 m_{er} Z r^8}. \end{aligned} \tag{12.50}$$

The authors of paper [18] wrote (we updated the numeration of equations and references):

'Next, we consider the orbits of the electron in the plane of the orbits of the muon and the nucleus, i.e., $\theta = \pi/2$. It is easy to check, by looking at the first and second derivatives of the effective potential from equation (12.50) with respect to θ, that this position corresponds to the stable equilibrium. In this case, the effective potential will be a Coulomb potential with two terms of the $1/r^n$-perturbation (from now on we will use the atomic units $e = m_e = \hbar = 1$, as in the overwhelming majority of theoretical papers in atomic physics):

$$U_{eff} = -\frac{Z-1}{r} + \frac{R_{\mu 0}^2}{4r^3} + \frac{9\varepsilon^2 m_{\mu r} R_{\mu 0}^4}{16 m_{er} Z r^8}. \tag{12.51}$$

The calculation of the $1/r^n$-perturbation for the Kepler problem can be found, e.g., in [19] (the treatment for the cases $n = 2$ and $n = 3$ can be found also in the textbook [20]).'

The perturbation $-\beta/r^{n+1}$ added to the Coulomb potential $-\alpha/r$ leads to the following periapsis advance of the orbit:

$$\delta\Phi = 2m\beta\frac{\partial}{\partial L}\left(\frac{1}{L}p^{1-n}\int_\pi^0 (1 + \varepsilon_e \cos\varphi)^{n-1} d\varphi\right). \tag{12.52}$$

In equation (12.52), L is the angular momentum, ε_e is the eccentricity of its orbit, and $p = L^2/(m\alpha)$; m is the reduced mass of the system 'Coulomb center-particle', the particle being the electron in our case. The periapsis advance divided by 2π yields the precession frequency in units of the Kepler frequency:

$$\frac{\omega_{pip}^{(1)}}{\omega_K} = -\frac{3}{4} m_{er}^2 (Z - 1) S^2. \tag{12.53}$$

In equation (12.53)

$$S = \frac{R_{\mu 0}}{L^2}. \tag{12.54}$$

The authors of paper [18] wrote (we updated the numeration of equations and references):

'The same result could be obtained by using equation (1.7.10) from book [16], where the potential corresponds to the gravitational potential of the oblate Earth and is mathematically equivalent to equation (12.13) (without the last term) with the following correspondence of the quantities:

$$GMm \Leftrightarrow (Z - 1)e^2,\ I_2 R^2 \Leftrightarrow \frac{2\gamma}{e^2},\ p \Leftrightarrow \frac{Z - 1}{2|E|},\ \gamma = \frac{e^2 R_{\mu 0}^2}{4(Z - 1)} \tag{12.55}$$

(the same treatment was used in paper [7]). Substituting the corresponding quantities into equation (1.7.10) from book [10] and considering the case of $\theta = \pi/2$, we obtain the same result as in equation (12.53) (with $e = 1$ in the atomic units).'

Figure 12.2 presents the dependence of the ratio $|\omega_{pip}^{(1)}/\omega_K|$ on $R_{\mu 0}$ in the case of $L = 3$ and $Z = 2$. One can see that the ratio remains small enough in this range of $R_{\mu 0}$, so that the validity condition of the analytical result for $\omega_{pip}^{(1)}/\omega_K$ from equation (12.53) is met.

Then the authors of paper [18] used equation (12.52) for treating the last term in equation (12.50) and got the next contribution to the scaled frequency of the precession as follows:

$$\frac{\omega_{pip}^{(2)}}{\omega_K} = -\frac{63\varepsilon^2 m_{\mu r} m_{er}^6 (Z - 1)^6 S^7 f(E_s)}{256Z},\ f(E_s) = 429 - 495E_s + 135E_s^2 - 5E_s^3. \tag{12.56}$$

In equation (12.56)

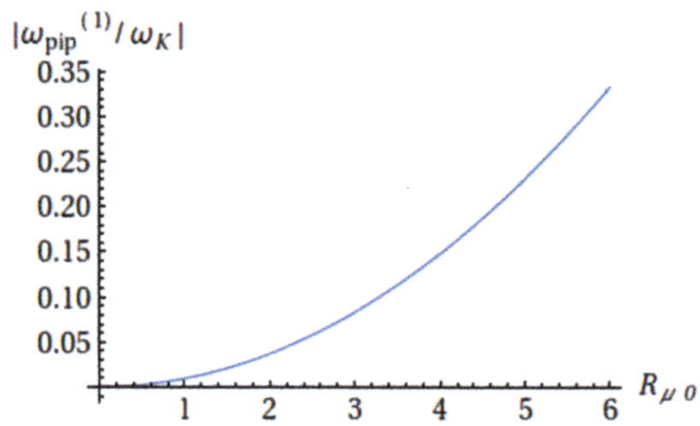

Figure 12.2. Plot of the ratio $|\omega_{pip}^{(1)}/\omega_K|$ from equation (12.17) versus $R_{\mu 0}$ for $L = 3$ and $Z = 2$. Reprinted from [18] with permission from the Open Journal of Microphysics.

$$E_s = \frac{2|E|L^2}{m_{er}(Z-1)^2} \tag{12.57}$$

is the scaled electron energy by its absolute value. It is related to the eccentricity of the electron orbit as follows:

$$\varepsilon_e = (1—E_s)^{1/2}. \tag{12.58}$$

For the bound motion one has

$$0 < E_s \leqslant 1. \tag{12.59}$$

We note that the quantity E_s is related to the semi-axes of the unperturbed ellipse: it is the square of the ratio of the semi-minor to semi-major axes.

Figure 12.3 depicts the dependence of the quantity f from equation (12.56) on E_s. The quantity f decreases from 429 to 64 as E_s grows from 0 to 1.

As an example, figure 12.4 shows the plot of the ratio $|\omega_{pip}{}^{(2)}/\omega_K|$ versus $R_{\mu 0}$ and ε for $E_s << 1$ (corresponding to the relatively large eccentricity of the electron orbit $(1-\varepsilon_e) << 1$), $L = 3$, and $Z = 2$ (so that $m_{\mu r} = 200.4$). It is seen that in these ranges of $R_{\mu 0}$ and ε, the additional precession frequency $\omega_{pip}{}^{(2)}$ remains sufficiently smaller than the Kepler frequency of the electron ω_K, which is the condition of the validity of the analytical result for $\omega_{pip}{}^{(2)}/\omega_K$ from equation (12.56).

The ratio $\omega_{pip}{}^{(2)}/\omega_{pip}{}^{(1)}$ (denoted below as K_{21}) of the additional contribution to the precession frequency from equation (12.56) to the primary contribution to the precession frequency from equation (12.53) is:

$$K_{21} = \frac{21\varepsilon^2 m_{\mu r} m_{er}^4 (Z-1)^5 S^5 f(E_s)}{64Z}. \tag{12.60}$$

Figure 12.5 shows the plot of K_{21} versus $S = R_{\mu 0}/L^2$ and E_s (defined in equation (12.57)) for $Z = 2$ and $\varepsilon = 0.02$.

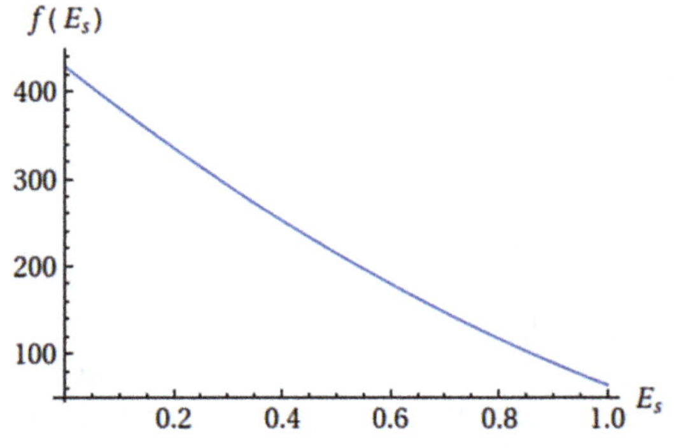

Figure 12.3. Plot of the function $f(E_s)$ from equation (12.56), where E_s is the absolute value of the scaled dimensionless energy of the electron defined in equation (12.57). Reprinted from [18] with permission from the Open Journal of Microphysics.

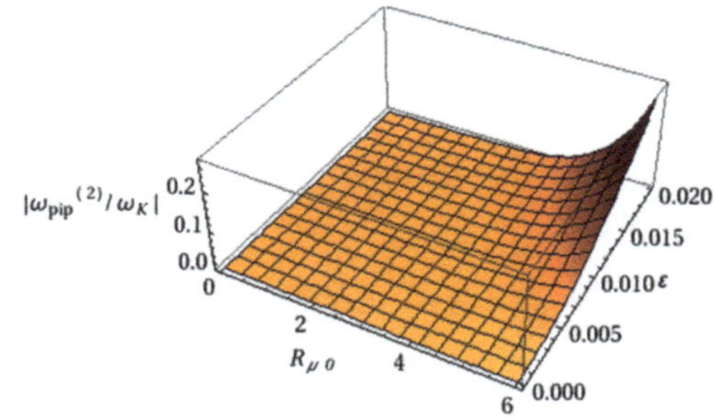

Figure 12.4. Plot of the ratio $|\omega_{pip}^{(2)}/\omega_K|$ from equation (12.56) versus $R_{\mu 0}$ and ε for $E_s << 1$ (corresponding to the relatively large eccentricity of the electron orbit $(1 - \varepsilon_e) << 1$), $L = 3$, and $Z = 2$ (so that $m_{\mu r} = 200.4$). Reprinted from [18] with permission from the Open Journal of Microphysics.

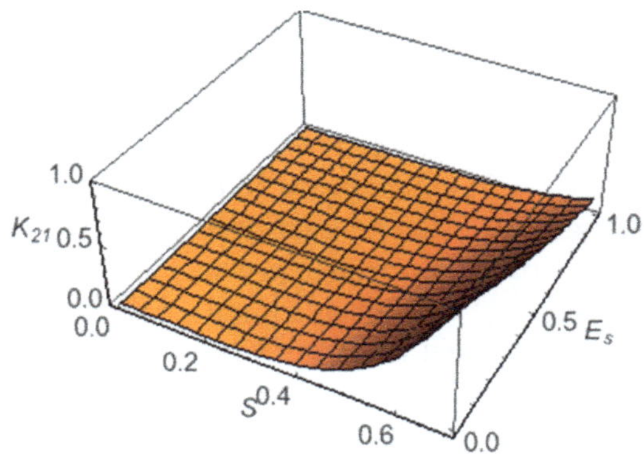

Figure 12.5. Plot of the ratio K_{21} of the additional contribution to the precession frequency from equation (12.56) to the primary contribution to the precession frequency from equation (12.53) versus $S = R_{\mu 0}/L^2$ and E_s (defined in equation (12.57)) for $Z = 2$ and $\varepsilon = 0.02$. Reprinted from [18] with permission from the Open Journal of Microphysics.

Figure 12.6 presents the plot of the ratio K_{21} versus S and Z for $\varepsilon = 0.02$ and $E_s \ll 1$.

Figure 12.7 demonstrates the plot of the ratio K_{21} versus $R_{\mu 0}$ and ε for $Z = 2$, $L = 3$, and $E_s \ll 1$.

From figures 12.5 to 12.7 one can see that the contribution $\omega_{pip}^{(2)}$ to the frequency of the precession due to by some eccentricity of the muonic and nucleonic orbits can become comparable to the main contribution (the main contribution being for the case of ε =0) within the ranges of the validity of the derived formulas for $\omega_{pip}^{(1)}/\omega_K$ and $\omega_{pip}^{(2)}/\omega_K$.

In summary, the authors of paper [18] generalized the results from the previous paper [7] by considering elliptical (rather than circular) orbits of the muon and the nucleus. They derived an additional contribution to the effective potential for the electron motion—the contribution depending on the eccentricity of the muonic and nucleonic orbits—assuming that the eccentricity was relatively small.

For the particular case of the planar geometry (i.e., where all the three particles move in the same plane), the authors of paper [18] obtained an additional contribution to the precession frequency of the orbit of the electron. They

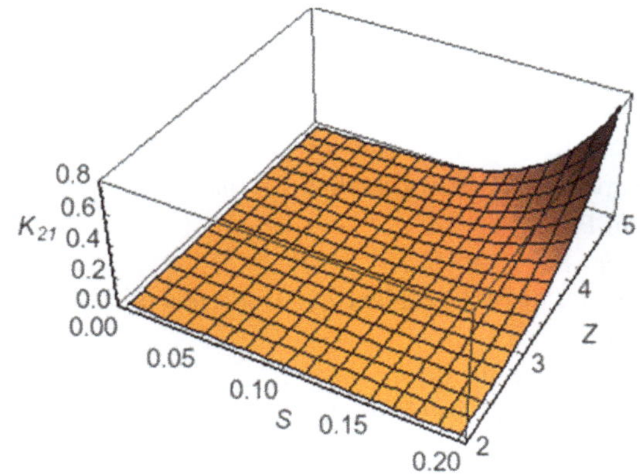

Figure 12.6. Plot of the ratio K_{21} versus S and Z for $\varepsilon = 0.02$ and $E_s << 1$ ($E_s << 1$ corresponds to a relatively large eccentricity of the unperturbed electron orbit). Reprinted from [18] with permission from the Open Journal of Microphysics.

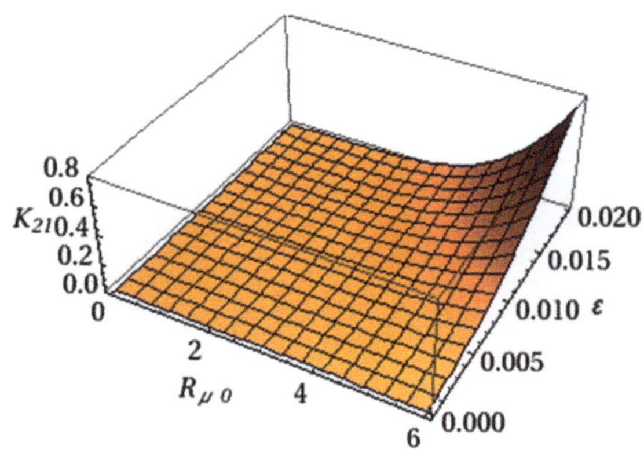

Figure 12.7. Plot of the ratio K_{21} versus $R_{\mu 0}$ and ε for $Z = 2$, $L = 3$, and $E_s << 1$. Reprinted from [18] with permission from the Open Journal of Microphysics.

demonstrated that the allowance for the eccentricity of the muonic and nucleonic orbits—even when the eccentricity is relatively small—can lead to significant changes of the precession frequency of the electronic orbit.

References

[1] Ponomarev L I 1990 *Contemp. Phys.* **31** 219

[2] Nagamine K 2001 *Hyperfine Interact.* **138** 5

[3] Nagamine K and Ponomarev L I 2003 *Nucl. Phys.* A **721** C863

[4] Chelkowsky C, Bandrauk A D and Corkum P B 2004 *Laser Phys.* **14** 473

[5] Guffin J, Nixon G, Javorsek D II, Colafrancesco S and Fischbach E 2002 *Phys. Rev.* D **66** 123508

[6] Oks E 2020 *Analytical Advances in Quantum and Celestial Mechanics: Separating Rapid and Slow Subsystems* (Bristol: IOP Publishing)

[7] Oks E 2020 *Can. J. Phys.* **98** 857

[8] Kapitza P L 1951 *Sov. Phys. JETP* **21** 588
[9] Kapitza P L 1951 *Uspekhi Fiz. Nauk* **44** 7
[10] Beletsky V V 2001 *Essays on the Motion of Celestial Bodies* section 1 (Basel: Birkhäuser) p 7
[11] Landau L D and Lifshitz E M 1965 *Quantum Mechanics* (Oxford: Pergamon)
[12] Bethe H A and Salpeter E E 1977 *Quantum Mechanics of One- And Two-Electron Atoms* (New York: Plenum Press)
[13] Oks E 2021 *Dynamics* **1** 1
[14] Kotkin G L and Serbo V G 1971 Problem 2.32 *Collection of Problems in Classical Mechanics* (Oxford: Pergamon)
[15] Oks E 2020 *New Astron.* **74** 101301
[16] Lisitsa V S 1987 *Sov. Phys. Uspekhi* **30** 927
[17] Lisitsa V S 1994 *Atoms in Plasmas* (Berlin: Springer)
[18] Kryukov N and Oks E 2021 *Open J. Microphys.* accepted
[19] Belitz D 2014 Lectures on Theoretical Mechanics p 95
[20] Landau L D and Lifshitz E M 1960 Problem 3 *Mechanics* (Oxford: Pergamon) ch 3

Appendix A

Analysis of the stability of the muonic motion in muonic-electronic quasimolecules based on a fully-stripped multicharged ion

The authors of paper [1] analyzed the stability of the motion of the muon in the μZe system—in addition to their primary results represented in chapter 10 of this book. For this purpose they wrote the corresponding Hamiltonian in the cylindrical coordinates as

$$H = \frac{1}{2}\left(p_\rho^2 + p_z^2 + \frac{p_\varphi^2}{\rho^2}\right) + U(z, \rho). \tag{A.1}$$

In equation (A.1), the muon potential energy is

$$U(z, \rho) = -\frac{Z}{\sqrt{z^2 + \rho^2}} - \frac{Z'}{\sqrt{(R - z)^2 + \rho^2}} \tag{A.2}$$

with $Z' = -1$. The Hamiltonian does not depend on the coordinate φ—consequently the corresponding conjugated momentum is conserved:

$$|p_\varphi| = \text{const} = L. \tag{A.3}$$

By combining equations (A.3) and (A.1), the authors of paper [1] separated the φ-motion from the z- and ρ-motions. For the latter they got the Hamiltonian

$$H_{z\rho} = \frac{1}{2}(p_\rho^2 + p_z^2) + U_{\text{eff}}(z, \rho) \tag{A.4}$$

with U_{eff} in scaled notations from equation (10.2) represented as

$$U_{\text{eff}} = -\frac{Z}{R}\varepsilon. \tag{A.5}$$

doi:10.1088/978-0-7503-3939-1ch13

In equation (A.5), the scaled muonic energy ε is:

$$\varepsilon = \frac{Z}{\sqrt{w^2 + p}} - \frac{1}{\sqrt{(1 - w)^2 + p}} - \frac{\ell^2}{2p}. \tag{A.6}$$

Then the authors of paper [1] equated to zero the partial derivatives of $\varepsilon(w, v)$ for obtaining the equilibrium conditions:

$$\frac{\partial \varepsilon}{\partial w} = 0, \ \frac{\partial \varepsilon}{\partial v} = 0. \tag{A.7}$$

The first relation in equation (A.6) yields:

$$p = w^{2/3}(w - 1)^{2/3} \frac{Z^{2/3}(w - 1)^{4/3} - w^{4/3}}{(w - 1)^{2/3} - Z^{2/3} w^{2/3}}. \tag{A.8}$$

The second relation in equation (A.6) yields:

$$\ell^2 = p^2 \left(\frac{Z}{(w^2 + p)^{3/2}} - \frac{1}{((1 - w)^2 + p)^{3/2}} \right). \tag{A.9}$$

Next, the authors of paper [1] analyzed small deviations from equilibrium

$$\delta w = w - w_0, \ \delta v = v - v_0. \tag{A.10}$$

In equation (A.10), w_0 is the scaled z-coordinate, at which the equilibrium values are calculated; later on, the subscript 0 will be omitted from w. The authors of paper [1] expanded ε in powers of these small deviations:

$$\varepsilon \approx \varepsilon_0 + \frac{1}{2} \frac{\partial^2 \varepsilon}{\partial w^2} \delta w^2 + \frac{1}{2} \frac{\partial^2 \varepsilon}{\partial v^2} \delta v^2 + \frac{\partial^2 \varepsilon}{\partial w \partial v} \delta w \delta v. \tag{A.11}$$

In equation (A.11), ε_0 is the equilibrium value of the scaled potential energy. The second derivatives calculated at the equilibrium are:

$$\frac{\partial^2 \varepsilon}{\partial w^2} = \frac{1}{(w^2 + v_0^2)^{3/2}} \left(\frac{1}{1 - w} - \frac{3wP}{Q^2} \right) \tag{A.12}$$

$$\frac{\partial^2 \varepsilon}{\partial v^2} = \frac{1}{(w^2 + v_0^2)^{3/2}} \left(\frac{1}{1 - w} + \frac{3wP}{Q^2} \right) \tag{A.13}$$

$$\frac{\partial^2 \varepsilon}{\partial w \partial v} = \frac{1}{(w^2 + v_0^2)^{3/2}} \frac{3wv_0(2w - 1)}{Q^2}. \tag{A.14}$$

The quantities P and Q are the following notations:

$$P = w(1 - w) + v_0^2, \quad Q = \sqrt{(w^2 + v_0^2)((1 - w)^2 + v_0^2)}. \tag{A.15}$$

The derivative from equation (A.14), generally speaking, differs from zero. Therefore, the authors of paper [1] rotated the reference frame to diagonalize the corresponding matrix:

$$\delta w' = \delta w \cos\alpha + \delta v \sin\alpha, \ \delta v' = -\delta w \sin\alpha + \delta v \cos\alpha. \tag{A.16}$$

In equation (A.16), α is the angle of the rotation. It turned out that the rotation by the following value of the angle α achieves the diagonalization:

$$tg2\alpha = 2\frac{\partial^2 \varepsilon}{\partial w \partial v}\left(\frac{\partial^2 \varepsilon}{\partial w^2} - \frac{\partial^2 \varepsilon}{\partial v^2}\right)^{-1} = \frac{(1 - 2w)v_0}{P}. \tag{A.17}$$

At this value of α, the expansion of ε can be represented as follows:

$$\varepsilon \approx \varepsilon_0 + \frac{1}{2}\delta w'^2 \omega_-^2 + \frac{1}{2}\delta v'^2 \omega_+^2. \tag{A.18}$$

In equation (A.18)

$$\omega_\pm = \sqrt{\frac{1}{(w^2 + v_0^2)^{3/2}}\left(\frac{1}{1 - w} \pm \frac{3w}{Q}\right)}. \tag{A.19}$$

The quantity ω_+ is real under the condition

$$\frac{1}{1 - w} + \frac{3w}{Q} \geqslant 0 \tag{A.20}$$

and ω_- is real under the condition

$$Q \geqslant 3w(1 - w). \tag{A.21}$$

Under the conditions (A.20) and (A.21), ω_+ and ω_- are the scaled frequencies of small oscillations in the direction of the normal coordinates $\delta v'$ and $\delta w'$, respectively. In this situation, the equilibrium is stable: at the equilibrium there is a two-dimensional minimum of the scaled potential energy.

The condition (A.21) is always fulfilled for any value of w from the allowed range given by equation (10.6). The condition (A.20) can be represented in the form:

$$(w^2 + v_0^2)((1 - w)^2 + v_0^2) - 9w^2(1 - w)^2 \geqslant 0. \tag{A.22}$$

By employing equations (10.4) and (10.7), the inequality (A.22) can be rewritten as follows:

$$Z^{2/3}(1 + 16\gamma^6 + \gamma^{12}) - 9\gamma^4(Z^{4/3} + \gamma^4) \geqslant 0. \tag{A.23}$$

In the allowed range of the parameter γ ($Z^{1/3} < \gamma < +\infty$), the inequality (A.23) is always satisfied. The final conclusion is that the muonic motion is indeed stable.

Reference

[1] Kryukov N and Oks E 2014 *Can. J. Phys.* **92** 1405

Appendix B

Rigorous value of the electric dipole moment of the atomic electron following from the separation of the center-of-mass motion and the relative motion in the three-body problem

The author of paper [1], the main results of which are represented in section 12.2 of this book, considered the system of three masses M, m_1, and m_2 located at $\mathbf{R}$, $\mathbf{R}_1$, and $\mathbf{R}_2$, respectively. It was assumed that $M >\max(m_1, m_2)$. The vectors were measured with respect to the origin located at the center of mass. Therefore:

$$\mathbf{R} + Mm_1\mathbf{R}_1 + m_2\mathbf{R}_2 = 0. \tag{B.1}$$

The relative coordinates of masses m_1 and m_2 are

$$\mathbf{r}_1 = \mathbf{R}_1 - \mathbf{R}, \quad \mathbf{r}_2 = \mathbf{R}_1 - \mathbf{R}. \tag{B.2}$$

Consequently:

$$\mathbf{R} = -(\mathrm{m}_1\mathbf{r}_1 + \mathrm{m}_2\mathbf{r}_2)/(\mathrm{M} + \mathrm{m}_1 + \mathrm{m}_2), \tag{B.3}$$

$$\mathbf{R}_1 = [(M + m_2)r_1 - m_2r_2]/(M + m_1 + m_2), \tag{B.4}$$

$$\mathbf{R}_2 = [(M + m_1)\mathbf{r}_2 - m_1\mathbf{r}_1]/(M + m_1 + m_2). \tag{B.5}$$

The kinetic energy T (as a part of the Lagrangian) is

$$T = (M\mathbf{V}^2 + m_1\mathbf{V}_1^2 + m_2\mathbf{V}_2^2)/2. \tag{B.6}$$

In equation (B.6), the time derivatives of vectors $\mathbf{R}$, $\mathbf{R}_1$, and $\mathbf{R}_2$, are denoted as $\mathbf{V}$, $\mathbf{V}_1$, and $\mathbf{V}_2$, respectively.

The author of paper [1] differentiated equations (B.3), (B.4), and (B.5) with respect to time and got:

$$T = m_{1r}\mathbf{v}_1^2/2 + m_{2r}\mathbf{v}_2^2/2 - m_1 m_2 \mathbf{v}_1 \mathbf{v}_2/(M + m_1 + m_2). \quad \text{(B.7)}$$

In equation (B.7), the time derivatives of vectors $\mathbf{r}_1$ and $\mathbf{r}_2$ are denoted as $\mathbf{v}_1$ and $\mathbf{v}_2$, respectively. In equation (B.7)

$$m_{1r} = m_1(M + m_2)/(M + m_1 + m_2) \quad \text{(B.8)}$$

and

$$m_{2r} = m_2(M + m_1)/(M + m_1 + m_2) \quad \text{(B.9)}$$

(these are the corresponding reduced masses).

Further, the author of paper [1] considered the situation where the object of the mass M has the charge Ze and the two other objects have the charge $-e$ (e being the electron charge by the absolute value). In this situation, the *instantaneous* dipole moment of the entire system of the three bodies is

$$\mathbf{d} = e(Z\mathbf{R} - \mathbf{R}_1 - \mathbf{R}_2). \quad \text{(B.10)}$$

By combining equations (A.3), (A.4), (A.5), and (A.10), the author of paper [1] got:

$$d = er_1[(Z - 1)m_1 + M + m_2]/(M + m_1 + m_2) + r_2[(Z - 1)m_2 + M + m_1] \\ /(M + m_1 + m_2). \quad \text{(B.11)}$$

For the case where the object having mass M is the nucleus of charge Z, the object having mass $m_1 = m_\mu$ is the muon, and the object of mass $m_2 = m_e$ is the electron, the averaged dipole moment can be obtained from the following considerations. The muon revolves in a circular orbit: therefore, the average value $< \mathbf{r}_1 >$ is zero. The electron revolves, generally speaking, in an elliptical orbit: therefore, $<\mathbf{r}_2>$ is not zero. Thus, the average dipole moment has the form:

$$<d> = e < r_2 > [(Z - 1)m_2 + M + m_1]/(M + m_1 + m_2). \quad \text{(B.12)}$$

Reference

[1] Oks E 2021 *Dynamics* **1** 1

www.ingramcontent.com/pod-product-compliance
Ingram Content Group UK Ltd.
Pitfield, Milton Keynes, MK11 3LW, UK
UKHW051849100926
8556IPUK00005B/538

9 780750 339377